Like There's No Tomorrow

Climate Crisis, Eco-Anxiety and God

— FRANCES WARD —

Sacristy Press

Sacristy Press
PO Box 612, Durham, DH1 9HT

www.sacristy.co.uk

First published in 2020 by Sacristy Press, Durham

Map illustrations by Kevin Sheehan of Manuscript Maps
www.manuscriptmaps.com

Sacristy Limited, registered in England & Wales, number 7565667

British Library Cataloguing-in-Publication Data
A catalogue record for the book is available from the British Library

ISBN 978-1-78959-088-3

For Nancy Stumpenhuson

Contents

1. Resilience, Relinquishment, Restoration, Reconciliation 1
2. Till in the Clouds she Sings. 19
3. Larkrise to Skipton. 35
4. Desire Runs Deep. 47
5. Ely and Prickwillow 61
6. A Lament for Creation. 75
7. Peterborough and the River Nene 89
8. Hawthorn Hedgerows Galore. 109
9. Brindley Country 125
10. Bridgewater to Wigan Pier 147
11. Up the Flight. 165
12. Botany Bay 177
13. Onwards to Pendle. 197
14. A Time of Deep Engagement 211
Postscript: Cumbria 231

Bibliography 237
Notes 240

CHAPTER 1

Resilience, Relinquishment, Restoration, Reconciliation

> O Lord, rebuke me not in your wrath;
> neither chasten me in your fierce anger.
> Have mercy on me, Lord, for I am weak;
> Lord, heal me, for my bones are racked.
> My soul also shakes with terror;
> how long, O Lord, how long?
>
> *Psalm 6:1–3*

If ever you've suffered from insomnia, Psalm 6 is for you. It's one of the "penitential psalms", supposedly written by King David as he is wracked by grief and guilt for taking Bathsheba from Uriah the Hittite, then causing Uriah's death on the battlefield.[1] Perhaps you've had the experience of dreaming again and again—the same dream, maybe. You've lost something: your purse, or some crucial papers for the meeting next day; a child in the supermarket. You wake early, and immediately your stomach sinks like a stone as acute anxiety hits hard. Something is wrong, and you are living with a deep dis-ease that cannot be resolved.

> Turn again, O Lord, and deliver my soul;
> save me for your loving mercy's sake.
> For in death no one remembers you;
> and who can give you thanks in the grave?
> I am weary with my groaning;
> every night I drench my pillow
> and flood my bed with my tears.
>
> *Psalm 6:4–6*

It couldn't go on. I was restless and unfocused at work. It was hard to get to the bottom of the anxiety that was exhausting me. Now I look back, and the factors were many, but the underlying one (which has outlasted all else) was a deep foreboding about the future of the planet.

It seemed that no one was taking it seriously enough. I couldn't enjoy work, and I knew I was close to stress. "I was weary with my groaning." I had lost the sense of the duty and joy that is at the heart of priesthood, a forward momentum that carries things on. Now it's called eco-anxiety—but it didn't have a name (at least as far as I knew) two or three short years ago, before the Greta effect.

I needed time to look square into that pit of misery I carried, which I was beginning to articulate for myself—that the way things are going, this earth might well be uninhabitable for our children's children. As someone who wonders about God, and seeks to live faithfully, this means profound questions:

> My soul also shakes with terror;
> how long, O Lord, how long?

Profound questions

Where is God if earth becomes uninhabitable? If humans become extinct because of climate change, who will praise the creator of all things?

Over recent months, Greta Thunberg has stirred the schoolchildren of the world with her quiet, undemonstrative passion. "Extinction Rebellion" has hit the headlines, campaigning for political action, initially with large numbers arrested, but then gathering global momentum with millions turning out on strike on designated days.[2] David Wallace-Wells is one of a number of writers publishing books on the future, with his prediction of the uninhabitability of the earth.[3] The work of Jem Bendell on "Deep Adaptation" has faced into climate catastrophe and given voice to its inevitability.[4] His paper has captured the imagination of many—as I write, over 400,000 have read the paper online, where it's gone viral. Many now realize it's no longer good enough to go softly, softly; that it might be too late anyway, as we face into catastrophe.

There is, however, little if any adequate theological engagement with the issues, although Hannah Malcolm is changing that with her win of the 2019 Theology SLAM prize,[5] entitled "Climate Chaos and Collective Grief".[6] Little else touches the deep lament and fear I feel as I contemplate the future as someone who holds Christian faith. It's difficult to write or talk about the possibility of what Bendell calls "inevitable near-term social collapse due to climate change". His purpose is to enable readers to take the opportunity to reassess their work and life, "like there's no tomorrow". He has promoted discussion in a scholarly environment that, on the whole, closes down worst-case scenarios because they are too bleak. He is building a global network of counsellors and thinkers now engaged, largely in response to his paper, in the psychological and political challenges that hit full on when you realize time is running out—fast. Theological voices need to join this movement.

Inside this experiment

In *Theology and Ecology Across the Disciplines*, Philip J. Sakimoto offers a prologue to aid understanding of the science of climate change.[7] He describes how so-called "greenhouse gases" (carbon dioxide, methane) trap heat in the earth's atmosphere. The more carbon dioxide there is, the warmer the planet will be. He writes:

> We are, at base, conducting for the first time an experiment in seeing how our planet responds to having copious amounts of greenhouse gases placed in its atmosphere. Since we are living *inside* this experiment, we cannot afford to wait until the experiment has played itself out to find out what the consequences might be.[8]

He continues to chart how the first warning that excess carbon dioxide might cause global warming came from John Tyndall in 1859, followed by Svante Arrhenius who predicted that the temperature could rise by five or six degrees Celsius—but this was unlikely to occur for three thousand years, at the then current rate of fossil fuel use. He could not have predicted just how fast subsequent generations would burn them. Before the industrial age, the carbon dioxide concentration never exceeded 300 parts per million—through eight orbital cycles of ice age and solar heating, over 800,000 years. In 2015, carbon dioxide concentrations passed 400 parts per million, and could exceed 600 parts per million by mid-century.[9]

Global warming is already happening, and the rate is increased by feedback loops: like those, for example, triggered by the melting of ice and snow, which then no longer reflect sunlight. Thawing tundra means vegetation decomposes, releasing methane into the atmosphere, which traps thirty times more heat than the same amount of carbon dioxide. Vegetation on land and phytoplankton in the ocean are responding to increasing warming by a decrease in the uptake of carbon dioxide, and a decrease in the production of oxygen. There are not the equivalent number of feedback loops acting to slow down the rate of global warming. Sakimoto says:

> Observations of the actual rate of global warming corroborate the sense that matters are getting seriously out of hand: the current rate of warming is significantly faster than it has been any time in the past, and the impacts of this warming are very clearly already upon us.[10]

Climate change brings changing weather, which will only become more extreme—storms, drought, wildfires. The spread of diseases is likely, as well as growing incidences of death from extreme heat, malnutrition, and water shortages. As living conditions deteriorate, there will be increased forced migration and civil conflicts, such that by 2050 the number of climate migrants may reach 200 million people, mainly children and the poor.[11]

We can't afford to delay. Sakimoto suggests:

> The essential question, therefore, is: *Do we have the societal will necessary to make the economic, social, regulatory, and energy infrastructure changes needed to convert to renewable energy?*[12]

He likens it to a world war, which requires a much more rapid and serious effort if we are going to win.

The chances are we won't.

Who gives thanks in the grave?

If you believe in God, there's a whole set of questions around this verse from Psalm 6:

> For in death no one remembers you;
> and who can give you thanks in the grave?

If the earth becomes uninhabitable, and humans become extinct (as some predict), is this God's will? Some Christians say so; I most certainly don't. Where is God, if there is no one to praise God? Does God begin again, waiting in eternity for human consciousness to arise to the extent that a response to God is possible (in a way not offered, as far as we usually suppose, by other creatures)? Christian theology says, rightly, that God is still God, even if there is no world. But the extinction of humanity with the rest of creation would leave the narrative structure of creation and redemption as the basis of God's relationship with humanity no longer meaningful, in any clear sense. Some argue that the world will regenerate over millions of years. But what hope does that hold for our children and grandchildren? What happens to faith, hope, and love if there is no longer a human response to God's grace?

Christian churches and theologians find it difficult to talk or preach or engage with the possibility of climate catastrophe and an uninhabitable earth, for the questions are enormous. Nevertheless, we need to struggle further to understand: because faith in God needs to engage with the reality of the tragic loss of creation through anthropogenic impact. It might not happen, but better to have anticipated the worst-case scenario as it unfolds; and if it doesn't, to give thanks in the awareness of what might have been.

Faith, if it is anything, keeps hope alive in the covenant of love that God established in the beginning, and renews in every living moment.

That covenant, though, doesn't guarantee a happy ending, so in what sense can we live with hope?

If we're living in the end times, then how should we live? Wracked with anxiety and guilt, with foreboding deep as death? How is it possible to live hopefully, even as we face realistically the inevitability of the radical impact of an unpredictable climate, rising sea levels, the collapse of biodiversity? How do we remain faithful to God and loving to our neighbour, particularly if our neighbours are exiles and immigrants because their homes are no longer inhabitable? What do we tell our children and grandchildren, so they don't grow up completely overwhelmed by anxiety, such that mental illness levels continue to soar?

This book is my personal attempt to think through some of these questions; to continue to have faith, hope, and love in response to God. I live with constant low-grade (and sometimes high-grade) anxiety, and I need to reflect theologically and find ways to pray and feel about the future that mean I can continue to live in hope. Not just ordinary hope, but a fierce hope that hopes against hope. I found myself seeking to turn the deep lament I felt into something fierce enough to mean I live life to the full, like there's no tomorrow.

Life-change

My husband Peter, after a long and successful career as a paediatrician, had responded to a call to be ordained and had gone to train at the College of the Resurrection in Mirfield, West Yorkshire. He was to serve his curacy in Workington, Cumbria. Inspired by his life-change, and also seeking to address that anxiety, it seemed a good time for me to leave full-time employment as well, after thirty years of ministry. During the early summer of 2018, I took our new narrowboat from March to Skipton. It was a journey through the heart of England; a slow, slow journey that gave me the space to think, feel, and pray. To attend to the world around and relish its history and natural beauty; to wonder what needs to be done for humanity to change its greedy, consumerist lifestyles. It was a voyage that changed my outlook fundamentally, from despair to hope, through lament; to find God's grace all around me, and in that grace to begin to hope, fiercely, with a deepened faith and love.

To respond to the inevitability of climate catastrophe is going to demand more and more from us over the months and years ahead. Political action is already late—too late, many believe. We need to stir politicians and institutions, economists, and everyone else to different priorities, and very quickly. There are many already engaged, and offering leadership, as the urgency of the disaster that is before us becomes more and more pressing. This will come to have an overwhelming impact on our lives at every level, and the sooner we respond the better. It helps to take time to appreciate deeply what it is we are fighting for. As the narrowboat made her way through the heart of England, the ordinariness of life became real and precious.

A contemplative spirit

It is clear to me that we need to balance the urge to immediate activism with a contemplative spirit—and it is in that spirit that this book is written. For if we are to continue with faith, hope, and love—particularly if we hold faith in God—we need to draw on that faith as a resource. We need to continue to be inspired by the God who creates, redeems, and sustains the world. This book tells the story of my deep engagement with the natural world around, as I took time to lament the passing of the normalities of the future I no longer think will happen. Activism is crucially important—to march on the streets as Extinction Rebellion do; changing the consumerist habits of our generation is essential, thereby resisting the market forces that commodify our lives. Any activism, though, needs to be inspired and sustained by a process that takes us, again and again, from despair, through lament, to fierce hope that has eyes to see and ears to hear the grace of God in all things, and to trust that grace.

◆ ◆ ◆

That narrowboat voyage didn't solve anything. It was a time fuelled by diesel, so deeply contradictory. It's perceived as a middle-class thing to do—though that perception isn't borne out: the canals now provide a home for many, many people of an extraordinary range of backgrounds. It might seem a real indulgence, to take off for six weeks, away from the need for action. I make no apology, though, for I'm convinced that if we're going to survive into the future, we need to develop the contemplative resources to support the inevitable activism that will be required to motivate politicians and our neighbours in the street to change in a radical direction, in terms of lifestyles.

The six weeks enabled me to become immersed in the world of the canal bank, the history of the English canal system, the people of today. I had time to rediscover the God who sustains and regenerates all life, as I entered deeply into the surrounding world during that month of May, as all around the natural world showed its generative power. The environment enfolded me and taught me how participation in God changes us, if we allow. I came to appreciate how everything created has its being in God in all its glorious particularity, with each being called to be what it is meant to be, in every moment. I came to ask of myself a question that emerged as fundamental: how can I be what I am meant to be, as I participate in the being of God? What of the fragile environments around? How do they have their being in God? How can human beings work with God to enable that being, rather than to change, or control, or destroy?

As I journeyed through the heart of England, I found myself exploring the history of the land and water we travelled through, with a deep appreciation of the way the past shapes the present. The canals were the result of the industrialization of England, and again and again I was reminded of how utilitarian the human approach has been—both in urban and rural contexts. If we are to explore (very late in the day) how the environment is meant to be within the great economy of God, we will ask very different questions, and work towards very different ends.

We need to feel God's wrath

My eyes are wasted with grief
and worn away because of all my enemies.
Depart from me, all you that do evil,
for the Lord has heard the voice of my weeping.
The Lord has heard my supplication;
the Lord will receive my prayer.
All my enemies shall be put to shame and confusion;
they shall suddenly turn back in their shame.

Psalm 6:7–10

With the psalmist I can—we all should—repent of our fault before God: how humanity has brought this wonderful planet to such desecration and destruction. We need to feel God's wrath; to feel the heavy rebuke of God's anger. That burden is one we need to carry. But instead of giving in to the enemies of hopelessness and helplessness, and extreme, deadly anxiety, we are called to live each moment to the full, and to seek to be, and to allow other things to live and move and have their being in God, in the knowledge that all being becomes real through participation in the all-encompassing and ultimate reality of God. This is to see the world as sacramental: that there is a loving and creative whole in which all things have their being, and in which each is held and regenerated. Sacramental, and therefore offering grounds for hope and regeneration. Grace breaks through, even as it pulsates throughout creation. When we are aware, with eyes to see and ears to hear, God is present; God's grace makes all things new.

"I would still plant my apple tree"

Hope is generated every time we turn towards the God who creates, seeking God's love and forgiveness, looking for signs of the grace that energizes each particle of existence. Hope springs eternal, when we align ourselves with God's creative power.

This means changing our perception and our ways of life. It means changing from utilitarian, instrumental approaches to the natural world around, and seeking to work in harmony, as God's co-creators, nurturing and loving alongside other beings in all their rich biodiversity—working to reverse the devastating collapse of species because of human impact.

Luther apparently said, "If I knew the world would end tomorrow, I would still plant my apple tree." That impulse is a good and true one: that even in the face of utter disaster and tragedy, we can still do things as if there is a tomorrow, trusting, hopeful in a love that transcends death. Planting an apple tree (or better, an orchard) offers a hosting environment for myriad other creatures to exist.

It might seem a drop in the ocean within the enormity of the climate collapse we face, but such action is a turning to the light of the Creator God; it is an action of hope and faith in God's loving, generative power. As Rowan Williams says:

> Just as—a difficult and painful analogy—the message of love texted to a partner or child by someone in a plane plummeting towards a crash is a moment of grace, even if there is no outcome that would make us "content" with the tragedy. Something terrible is simply going to happen, unavoidably—but there is a turning towards light that is still open.[13]

This is to recognize that we continue to live in hope because, ultimately, the desire for life and love is stronger than despair. This is easier to say than to do, though, when often hope is a fragile and vulnerable thing, even seemingly useless in the face of catastrophe. But hope is not self-generating for the Christian, because she will affirm that God is faithful first—"The one who calls you is faithful" (1 Thessalonians 5:24). God initiates from before all time—in creation, and in the sustaining power of love that holds all things, even as they change and decay. Even as it seems the whole of creation plummets towards catastrophe, God does not give up on the wonderful creation that God continues to create. Even if we are filled with dreadful hopelessness.

To focus on our own hopelessness is to miss the meaning of God. For if faith means anything, it sustains hope that the Creator God doesn't just walk away, but is committed, absolutely, to creation—even if we don't know what that commitment might look like down the line. Nothing can separate us from the love of God in Christ Jesus . . . not even climate catastrophe. Which isn't a recipe for sitting back and doing nothing. Again, Rowan Williams:

> Hope is a condition here and now; not a foretelling of what lies ahead, and certainly not an inoculation against loss, but the affirmation that God is bound to the finite reality that God loves, and it is God's business to honour that binding. Meanwhile, we keep ourselves open to the God who acts and speaks in the present, and we labour at whatever we can do to prevent catastrophe, even if we are fearful that all our effort is too late. We still celebrate the swallow in the moment: anticipating loss can be a way of slipping away from a gift that is here and now. Without the celebration now, the loss would actually be softened, in a strange way. The risk of loss gives the joy a kind of fierceness.[14]

This book is my attempt to find a fierce hope by which I can live and inspire others to live. To reach that fierce hope I need to lament—to allow that ancient practice to find expression and turn the debilitating fear and anxiety outwards into action. Action that is deeply rooted in a contemplative spirit that knows dread and transforms it, through faith in God.

It's a similar process to that described by Jem Bendell, who writes for audiences around the world who are waking up to the enormity of what faces the planet, and who need resources to continue to live without despair. He commends a "Deep Adaptation Agenda" to help guide discussions on what we might do once we recognize climate change is an unfolding tragedy. He outlines his twenty-five-year engagement with the research, and says:

> Environmental scientists are now describing our current era as the sixth mass extinction event in the history of planet Earth, with this one caused by us. About half of all plants and animal species in the world's most biodiverse places are at risk of extinction

> due to climate change (WWF, 2018). The World Bank reported in 2018 that countries needed to prepare for over 100 million internally displaced people due to the effects of climate change (Rigaud et al., 2018), in addition to millions of international refugees.[15]

He offers a psychological assessment for why there is denial and resistance to working with the worst-case scenarios. He notes how neo-liberalism has shaped the agenda for decades now—not only economically, but also that of universities and research (and, I would add, the Church):

> The West's response to environmental issues has been restricted by the dominance of neoliberal economics since the 1970s. That led to hyper-individualist, market fundamentalist, incremental and atomistic approaches. By hyper-individualist, I mean a focus on individual action as consumers, switching light bulbs or buying sustainable furniture, rather than promoting political action as engaged citizens. By market fundamentalist, I mean a focus on market mechanisms like the complex, costly and largely useless carbon cap and trade systems, rather than exploring what more government intervention could achieve. By incremental, I mean a focus on celebrating small steps forward such as a company publishing a sustainability report, rather than strategies designed for a speed and scale of change suggested by the science. By atomistic, I mean a focus on seeing climate action as a separate issue from the governance of markets, finance and banking, rather than exploring what kind of economic system could permit or enable sustainability.
>
> This ideology has now influenced the workloads and priorities of academics in most universities, which restricts how we can respond to the climate tragedy.[16]

This analysis is correct, I believe—and captures just how difficult it is to find the political motivation required to make changes that should have begun to happen in the 1970s. With the exception of the excellent 2015 encyclical of Pope Francis, *Laudato si'*,[17] the Church (by which I mean all the main churches and denominations) has not been astute in its prophetic task, and has largely bought into the language and ideology of late twentieth-century neo-liberalism, with increasing emphasis on managerial approaches, the rhetoric of "growth", and the requirement to market everything, including holiness.

Bendell concludes:

> It is a truism that we do not know what the future will be. But we can see trends. We do not know if the power of human ingenuity will help sufficiently to change the environmental trajectory we are on. Unfortunately, the recent years of innovation, investment and patenting indicate how human ingenuity has increasingly been channelled into consumerism and financial engineering. We might pray for time. But the evidence before us suggests that we are set for disruptive and uncontrollable levels of climate change, bringing starvation, destruction, migration, disease and war.[18]

He offers his "Deep Adaptation Agenda" to those who recognize the reality of climate collapse, and who don't want to give up in despair. Firstly, he advocates developing resilience. He quotes the explanation of the Stockholm Resilience Centre (2015) that "resilience is the capacity of a system, be it an individual, a forest, a city or an economy, to deal with change and continue to develop. It is about how humans and nature can use shocks and disturbances like a financial crisis or climate change to spur renewal and innovative thinking."[19] Resilience draws on observation of ecosystems that overcome disturbance by increasing their complexity.

Next, he outlines two more "Rs", Relinquishment and Restoration:

> [The need for more than resilience] brings us to a second area of this agenda, which I have named "relinquishment." It involves people and communities letting go of certain assets, behaviours and beliefs where retaining them could make matters worse. Examples include withdrawing from coastlines, shutting down vulnerable industrial facilities, or giving up expectations for certain types of consumption. The third area can be called "restoration." It involves people and communities rediscovering

> attitudes and approaches to life and organisation that our hydrocarbon-fuelled civilisation eroded. Examples include re-wilding landscapes, so they provide more ecological benefits and require less management, changing diets back to match the seasons, rediscovering non-electronically powered forms of play, and increased community-level productivity and support.[20]

Resilience, Relinquishment, Restoration.

Reconciliation is there, too, a fourth "R"—in work that Bendell offers, with others, to increase awareness and training.[21]

A theological response to the tragedy

As I look back one year on from the narrowboat voyage from March to Peterborough, then onwards in the north-westerly direction to Wigan, through Blackburn, and over the Pennines to Skipton, I wish I'd known Bendell's "deep adaptation". It would have helped enormously as I found myself responding to a profound desire to re-orientate my life. As it is, I can revisit what I wrote then, in journal and blog, and reframe the experience in the light of what I now know, drawing on the resources of faith, principally the penitential psalms (but not only those), to offer a theological response to the tragedy that unfolds before us. My instinct was right—to seek a deep and slow engagement and participation in the world of nature. I took a sabbatical, which is to sink deeply into the sabbath, the time of rest and refreshment, in order to review and reinvigorate oneself for the resilience, relinquishment, reconciliation, and restoration that is required.

It enabled me to find a place of fierce hope within me by living in the moment—every moment—and finding the grace of God there, in a restorative way. Through deep dread and lament the light and grace emerged that grew resilience and response in me, resourced by the Christian faith, to renew my love of God's world, of neighbour, with a generous, open spirit. Instead of dark despair, I could cherish and celebrate, and share the fierce hope of others, expressed, for example, as Charlie Burrell and Isabel Tree do by rewilding through the Knepp

Wildland Project.[22] I began to examine all the habits and behaviours that had come to feel normal but which destroy and need to be relinquished. I remember a time when there was ice on the inside of my windows in winter. Now my home is heated to a ridiculous degree. What else, apart from over-heated homes, will we need to relinquish? How do we decide what to give up? Will politicians make it illegal to own a car that isn't electric? Could I do without a car altogether? How will we relinquish our dependence on fossil fuels with a target of zero by 2050 (when really, we can't wait that long)? That internal journey from despair to fierce hope is a necessary one for all who want to change themselves and encourage others to change too. It's an internal journey that needs to be travelled again and again, fuelled by a deep contemplation of the grace of God in our lives and in the world.

◆ ◆ ◆

Restoration. I know the narrowboat is fuelled by diesel. I know that's not good. But nevertheless, the six weeks I spent that summer were restorative, in all sorts of ways. The canals of England tell a story that is worth hearing—a story of thrusting and innovative industrialization and the development of manufacture and commerce. The canals were first fuelled by horsepower—real horse power. As our civilization came to depend on coal and steam, and then the internal combustion engine, the canals were overtaken: first by rail, then by road, leading to their disuse and dereliction in many areas. The canals have now, largely, come back to life as people use them for deep, slow leisure.

You can meet all England on narrowboats. You'd be surprised how many people of all backgrounds live aboard now—seeking a slower life—or hire boats for holidays, bringing the canals back to life after they had fallen into disuse when no longer viable as commercial routes. Tom Rolt's book *Narrow Boat*, first published in 1944,[23] led to the foundation of the Inland Waterways Association,[24] the beginning of revival for a different purpose. The canals are now corridors of wildlife and were glorious in May 2018. Dieter Helm commends such corridors in his *Green and Prosperous Land* for the way they snake through fields and cities, providing green connection amid the sterility of over-farmed regions and the desolation and post-industrial neglect of urban areas. Canals offer a history so easily missed by car. My voyage turned into an odyssey through restored waterways, now with potential to be so much more amidst the dereliction of country and town.[25]

It helped enormously to slow down and learn new skills that took me into my body. Over those six weeks I developed the instincts that come with training: the boat taught me how she would respond in the different conditions we encountered. I became skilled at manoeuvring through tight spaces; coping with busy town centres, deep locks, shower curtains entangled in the propeller. I became stronger and fitter. The deep anxiety was there, still, and would overwhelm in a moment as I lamented how few swallows there were, or counted wild flowers—pitiful numbers compared with Portugal, for instance, which has not suffered the intense farming practices the UK has been subjected to since the 1970s—or noted the lack of insects. Such sudden angst would set off a train of panic rising, which, as I drove along, would lead towards God. I learned to give thanks; to resist the anxiety, not to give in to its corrosion. The physicality helped enormously in the battle against the dark, bleak depression within. It held me in my body, and in the world around, instead of escaping into any number of fears and dark places.

Jem Bendell's work on deep adaptation offers an agenda I've been waiting for. It's time the churches were on board with this sort of analysis. Because the deep anxieties we all live with are compounded by faith. It's really hard to see the way through to a sense of hope for the future if the worst-case scenarios are inevitable.

Bendell writes:

> Human extinction and the topic of eschatology, or the end of the world, is something that has been discussed in various academic disciplines, as you might expect. In theology it has been widely discussed.[26]

But I'm not sure it has been, at all, in the light of the imminent end of the world. Bendell's "Deep Adaptation Agenda" of resilience, relinquishment, reconciliation, and restoration offers a way to voice and cope with the extreme anxiety that many people feel. It takes a scientist to remind the

Church of what it should be focused on. Bendell's idea of restoration suggests the need to dig deep into theological resources around loss and lament, in order to learn habits that can contain the anxiety and turn it to the good. How do we live with a fierce hope that can transform life by working actively for a viable future for the world and all its inhabitants?

> Turn again, O Lord, and deliver my soul; save me for your loving mercy's sake.

The Judaeo-Christian tradition offers ways to contemplate loss and can show how to lament. One way to stay with loss and lament is to turn, and turn again, to the profound emotional wisdom of the psalms.

A fierce and hopeful lament

This book is a fierce and hopeful lament for the future, written in the faith that God does not abandon the creation God brings into being. God is faithful, and calls from us a faithful response, even in the face of hideous disaster. So now I see a swallow and refuse the bleak anxiety, choosing instead to go deep and slow into the contemplation of life and a sense of thankfulness, which is to turn to the God who is life. God's life and love restores each created being to what it is meant to be, even to eternity. That deep knowledge gives resilience with, in addition, the ability to give up that which is not necessary to life.

One of the ways to go deep and slow is by drawing on our rich cultural and literary history. The psalms are fundamental to that heritage—looking into the face of personal loss and disaster, and enabling a response to tragedy. That summer I allowed the natural world around me to speak of hope as it reminded me of loss, and so to have a restorative effect as I learned habits of living more simply, and of travelling light, in a wonderful boat that weighs eleven tons, yet moves slowly and gently through the water that upholds it.

As we journeyed on, the countryside brought me renewed hope in the resilience of nature as I found ways in literature and reading to celebrate with a fierce hope the world around, God's eternally renewing creation.

We are called by God into being; we are called to live as we are meant to be, each in our own particularity. To slow down and attend to the life that pulsates in each tree ablaze in flower, or in the sight and sound of swifts screaming overhead, is to connect deeply with the love that will not let us go.

CHAPTER 2

Till in the Clouds she Sings

The Lark Ascending was her name. Fifty-seven-foot-long, our new narrowboat was ready. In April 2018 she was the 123rd boat built by Fox Narrowboats in March, near Peterborough.[27] Fox Narrowboats was set up by Charlie Fox, who died in 2012. He'd built the marina from scratch, leaving a family business that thrives, hiring out day boats and narrowboats to those who want to enjoy the Fenland rivers. These days, they build one boat to sell per year, so we had waited through the winter months, encouraged by progress reports and photos, as the hull was assembled from sheet steel, and the interior was fitted out to our specifications.

Her bottom plate is a centimetre thick, with side plates six millimetres and the superstructure four millimetres. She has a diesel Beta Marine Engine, gas for the cooker, solar panels, and a wood burner, which also heats the radiators. The three batteries are charged by the engine and the solar panels, and feed the 2.5kW inverter. There's a shower room, with a loo that flushes into a holding tank. A double bunk up towards the bow, with two single bunks that can be pushed together, makes it very comfortable for two adults.

All you'd need to live aboard.

Such a heavy thing. Eleven tons. Water-bound. Slow. So unlike the "Lark Ascending" of George Meredith's conception:

> For singing till his heaven fills,
> 'Tis love of earth that he instils,
> And ever winging up and up,
> Our valley is his golden cup,
> And he the wine which overflows
> To lift us with him as he goes.[28]

The poem inspired Ralph Vaughan Williams, who composed the justifiably popular classical piece with the same title which, when exquisitely performed, strings out those thin, high sounds as if they would break, high, high above the earth; then tumbling down, dropping, dropping, returning to the nest. The sound of the violin, tenuous, stretched, just short of breaking point, is the sound of nature, skylark above waste ground or estuary, captured by human art. A note of hope.

John Clare captured the same exquisite moment as he looked up, his hand to the plough, and saw a skylark far overhead:

> O'er her half-formed nest, with happy wings
> Winnows the air, till in the clouds she sings,
> Then hangs a dust-spot in the sunny skies,
> And drops, and drops, till in her nest she lies.[29]

The Lark Ascending: a narrowboat named to recall the power of poetry and music.

Disturbingly, during May that year we didn't hear a skylark at all. We lived with the lack, yearned for it in hope, and were disappointed as we travelled the waterways of England.

Disguising a deeper distress

The story of *The Lark Ascending* that summer took us through the canals and rivers from south-east to north-west, through towns and countryside cultivated and created by human beings for centuries. It was the early summer of 2018, and England was in cultural and political turmoil, trying to work out what it would mean to leave the European Union. As we travelled from March to Skipton at three miles an hour, it was hard to ignore the profound divisions and bitterness around, though most people we met didn't "go there", unwilling to discuss freely the Referendum of 2016. My sense was that it was unreal, disguising a deeper distress. It seemed a cloud had descended on the country—an acrid cloud that tasted sour in the mouth and throat, and left you speechless, unable to ask or talk about Brexit, except when with those who agreed with you. It was unspeakable, but not as unspeakable as that deeper matter beneath articulation. The cloud was real, too, made up of particulates, sulphur dioxide, nitrogen oxides, ground-level ozone, carbon monoxide, black carbon, lead, and methane.[30]

◆ ◆ ◆

I'd lived and worked, through my life, in many parts of the UK—in Cambridgeshire, Scotland, the East End of London, Lancashire, Manchester, Bradford, Suffolk—and now was about to move to Workington, Cumbria. I'd experienced different regions and aspects of England (and Scotland in the 1980s, as a student) and loved its hybridity and differences—the people who have come over the centuries and made their home here. I loved our political systems—the way they have been forged over centuries, and carry forward a constitutional and representative democratic polity that is second to none in the world, but now under real pressure from populism on right and left. I loved the history of England, its poetry and cultural adventurousness. I knew its fallibilities and failures too—that post-colonial legacy of Empire that left deep scars and contributed to a conflicted sense of nationhood. I was under no illusion about England and its flaws. It was under a bitter cloud, seeking ways to re-imagine or even re-romanticize itself. The soul of the nation needed to find a way forward that wasn't the nostalgic fantasy which, on the lips of some Brexiteers, sounded convincing, but was a chimera.

Somehow, the soul of the nation needed to know itself again: generous-spirited, confident, imaginative, cultured, diverse, and adventurous. I wanted to feel at home again in England—an England that was realistic about the enormity of the challenge of climate catastrophe. I wanted to be hopeful again, not dismayed at how the more profound and urgent issues of the future inhabitability of the planet were ignored as the nation tore itself apart. Where was the political action needed—now!—to reverse anthropogenic climate chaos? My intense frustration and restlessness required me to go deep and slow; to go to the heart of things, to draw on the history, literature, poetry, and theology of this land, as we travelled the waterways.

Poetry, metaphor, and the imagination

As a priest in the Church of England, to feel at home is to know God's presence in the world. This book belongs within a way of thinking and writing which is sometimes called "theopoetics". Theopoetics is a way to talk of God with poetry, metaphor, and the imagination. May and June 2018 were remarkable for cloudless skies; but I was in a cloud. I couldn't experience anything without a profound sense of loss. I was lamenting the future. Everything I did, felt, or thought was prefaced with, "like there's no tomorrow". This way of thinking could be seen, theologically, as a kind of *via negativa*, as a cloud of unknowing.

"The cloud of unknowing"—the title of a text of English mysticism from the twelfth century—suggests that God begins to be known when we enter into the darkness of despair and fear, of guilt and shame; when we reach the limits of human understanding and speech, and know ourselves in a cloud of obscurity. It seemed the cloud of unknowing surrounded me; it surrounded our troubled nation, our troubled world. I needed to be clothed in that cloud, to participate in it, seeking it as the mantle of God's grace, rather than running away into frenetic busyness, or angry displacement, or worse still, despair and debilitating anxiety. The cloud offered the opportunity to know again my dependence on the God of grace.

I was on a voyage—imaginatively, physically, psychologically, and spiritually—seeking a more truthful way of living, personally, into the future. Poetry and the imagination were crucial. Keefe-Perry's book on theopoetics, *Way to Water*, referred (to my delight) to someone I hadn't read for ages: Rubem Alves didn't just write *about* theopoetics, but "flung himself headlong into an attempt to realize the kind of integrated theological and poetic language that [others] wrote about."[31] Alves showed how theopoetics enables "a connection to a prophetic and poetic method and way into that which is eternal".[32] I looked to throw myself into ways of theological reflection that would help me into a way of hope rather than denial, and not just as a head exercise, but as total physical immersion into the eternity of God.

The Alves classic *The Poet, The Warrior, The Prophet* is the published version of the Edward Cadbury Lectures of 1990. His audience sat spellbound as he took them on a journey into the practice of unlearning and the wisdom of poetry.

Understanding attention

Another book that went with me was Iain McGilchrist's *The Master and his Emissary.*[33] McGilchrist has written, I reckon, one of the most significant books of the first decades of the twenty-first century. He offers an intriguing framework for understanding attention. He explains how the left and right hemispheres of the brain enable the person to attend differently to the world. His thinking develops in a metaphorical direction to explore how, increasingly, our culture is dominated by a left-hemispherical focus on detail, process, and instrumentality at the cost of the right hemisphere's encounter with the whole.

With his experience as a psychiatrist, McGilchrist writes convincingly about the pathologies of a society where the left hemisphere has the ascendancy, and we have lost what the right hemisphere seeks. We need what the right hemisphere attends to—a sense of the whole, engaging with the bigger picture—not only to reconcile the divisions, anger, and bitterness that tore at the heart of England, but also to face the future with hope. In four or five chilling pages, McGilchrist asks, "What would the left hemisphere's world look like?"[34]

In such a world, says McGilchrist:

> Cultural history and tradition, and what can be learnt from the past, would be confidently dismissed in preparation for the systematic society of the future, put together by human will. The body would come to be viewed as a machine, and the natural world as a heap of resource to be exploited. Wild and unrepresented nature, nature not managed and submitted to rational exploitation for science or the "leisure industry", would be seen as a threat, and consequently brought under bureaucratic control as fast as possible. Language would become . . . devoid of any richness of meaning, . . . suggesting a mechanistic world, . . . [with no overall feel for its metaphorical qualities].[35]

The journey I undertook was a voyage to discover again a vision and reality that could offer an alternative to the systematization, management, rational exploitation, bureaucratic control that I saw all around. We were caught, as a society and as a Church, in ways of living and working that destroyed the sacred canopy of wonder and awe which enables us to transcend our current attention to processes, systems, and detail into something whole. What might it be to attend to the world in a right-hemisphere way? Seeing the whole, and each part participating in that whole, as God means things to be?

This is what I love

There's an ancient prayer used at the altar, just before the Eucharistic Prayer begins.[36] It speaks of fragmented bits coming together, of the human contribution to making something wholesome from the natural world, to contribute to the realm of God. The words convey the sense that the world bears the creative power of God and is therefore sacramental, a visible sign of God's invisible grace. Attending to the physical reality around is to attend to God as behind, beneath, beyond all that is, in the materiality of things, actions, senses. It is to participate in the reality of God's loving creation, not to objectify it.

Augustine gives a good example of this sense of God's active love, or grace, as he responds to his own question in his *Confessions*:

> But what do I love when I love my God? Not material beauty or beauty of a temporal order, not the brilliance of earthly light; not the sweet melody of harmony and song; not the fragrance of flowers, perfumes, and spices; not manna or honey; not limbs such as the body delights to embrace. It is not these that I love when I love my God. And yet, when I love him, it is true that I love a light of a certain kind, a voice, a perfume, a food, an embrace; but they are of the kind that I love in my inner self, when my soul is bathed in light that is not bound by space; when it listens to sound that never dies away; when it breathes fragrance that is not borne away on the wind; when it tastes food that is

> never consumed by the eating; when it clings to an embrace from which it is not severed by fulfilment of desire. This is what I love when I love my God.[37]

God's active love in all things—all matter, our bodies, the natural world around—means we know our voice, perfume, food, and embraces hint at something else, even as they stay what they are. We know the beauty of the world around, noticing the brilliance of light, the sweetness of song, spice, honey, and we also know them in our imaginations, transformed by the love of God into the quintessence of materiality that transcends rationality and sense. Augustine brings us to the limits of our knowledge, to that which cannot be spoken—the apophatic, the ineffable. Like music, we see and use the physical around us (instruments, sheet of music, chair, stand) to create something that engages our left and right hemispheres simultaneously, to use McGilchrist's insights. Something other comes to being that has its own life yet doesn't leave behind the matter that is its medium.

All creation sings God's glory in this way. The matter of steel hull sings God's glory, just as the small, soaring bird does. All is inspired, alive, because it participates in the reality of God, and is sustained and created, continuously, by that love. The psalmist knew this:

> The heavens are telling the glory of God;
> and the firmament proclaims his handiwork.
> One day pours out its song to another
> and one night unfolds knowledge to another.
> They have neither speech nor language
> and their voices are not heard,
> Yet their sound has gone out into all lands
> and their words to the ends of the world.
>
> *Psalm 19:1–4*

As we voyaged throughout England in *The Lark Ascending*, I was seeking to know God more deeply in and from the world around. Despite the cloud that surrounded me, or rather, within the cloud itself, transforming its toxicity, the song of the grace of God would, I hoped, pour out to the ends of the world.

Entanglement

There are increasing numbers of minke whales, humpback whales, porpoises, and dolphins now swimming the coasts of the UK. All have been found snared in fishing gear in recent years. Porpoises are most likely to be caught by bottom-set trawling, dolphins by mid-water trawling. It's called entanglement. In Scotland, minke whales represent 87 per cent of reported entanglements. Humpbacks have been found over the last few years, tangled up and dying, with fishing nets and lines restricting the movement of these giant ocean creatures, or pinning their mouths shut.[38]

Entanglement.

It's a word that can go in many directions, literally and metaphorically. It's a word that captures how entangled we can feel, hopelessly and helplessly, in political and economic structures and systems that seem too complicated and self-serving to respond to the urgency of climate change.

In some ways, the clear division of Leave or Remain cut through the complexity of political and economic life in Britain today: understandable, when everything is so hard and seemingly hopeless, with growing polarization between rich and poor in our nation. Entangled in life, I found myself yearning for a renewed re-romanticization of the nation. I wanted the deep divisions of Brexit to heal, so we could find the way, as a nation, to be at unity, with a generous soul ready to play its part in a global world. A nation that could talk carefully and rationally about immigration with a welcoming spirit, rather than the giving of permission for the rise in the expression of prejudice against difference—Jewish people had recorded increased anti-Semitism; Muslim women feared to be on the streets in the hijab. A nation that could draw together to focus on improving the public goods—of education, health, travel, land and water use—with compassion for those who have suffered years of "austerity", biting deep. A nation that anticipated the impact of artificial intelligence on employment, and how automation would change our way of life. A nation that could take seriously the future, given the climate catastrophe that looms.

Brexit was the yearning for a safer world. It also expressed populist outrage against an establishment that was out of touch; with grievances born, as commentators said, of growing inequality over recent decades, amplified by the financial crash of 2008. We were living in a world that saw rising levels of unpredictability and threat: of power realignments, violence, and war. The environment was struggling, with climate change affecting all of us, and especially the poorest communities of the world.

Martin Ford says that when these differences come together:

> [T]he frightening reality is that if we don't recognize and adapt to the implications of advancing technology, we may face the prospect of a "perfect storm" where the impacts from soaring inequality, technological unemployment, and climate change unfold roughly in parallel, and in some ways amplify and reinforce each other.[39]

I felt—feel—entangled in forces and changes that threaten our very existence in the world of today. The cloud isn't just a cloud of unknowing: it's full of nightmarish traps and terrifying nets that stop us acting, protesting, engaging politically. Our politicians are trapped in nets too: unable to deliberate and decide what's best for the nation. As a Christian, I couldn't work out how to process all this. How to live in a society that was devastatingly divided and bitter, with deep fault lines, leaving us with political systems preoccupied to the detriment of other, crucial issues, including proper attention to getting universal credit right, and ensuring the NHS has a future, but with the challenge of climate chaos at the very top of the agenda. I wanted to be able to hold out thoughtful ways forward, ways that commended God, faith, hope, and love. Instead I felt at a loss, entangled in stuff that caught me into minutiae, management, trivial concerns, left-hemisphere processes and procedures that ensured I couldn't focus on the bigger picture. My hands—and my soul, mind, heart, and strength—were tied.

The softly-softly approach isn't working

For a while now, the line that has been taken—in public discourse, in church circles—is, "Don't be alarmist!" For fear and anxiety only demotivate. That approach must change. What we face is so threatening that it can no longer be ignored. The softly-softly approach isn't working. It's not one taken by *The Economist*, where I turn for honest, true news. Nor by David Wallace-Wells, whose book *The Uninhabitable Earth* is designed to inform and to stir humanity out of complacency. There are no Plan Bs when it comes to climate change. There is evidence that climate change has already changed our "normal" beyond return.

I'm profoundly scared. I'm scared for my potential grandchildren and their ordinary desires—what work they'll do, who they'll marry, their children—all that ordinary stuff. What we face will undermine all those desires. If climate change continues as its trajectory suggests, our whole civilization is under threat.

We face even greater migration of peoples, refugees of climate change. We face widespread destruction of species, starting with the insects, but not ending there. We face the warming and acidification of the seas, and the melting—the complete melting—of the polar caps, with rising sea levels and further disruption of weather patterns. We are beginning to see the unpredictability of what lies ahead as we watch Australia's temperatures soar and Chicago colder than parts of Mars. Climate changes will escalate.

One of the real demotivators to change is a sense of hopelessness at the global scale of the problem. What can I do, entangled in my day-to-day life? We read that, despite the Paris Agreement, recent renewed commitments, gathering concern, and political will, the demand for oil is rising. Wallace-Wells says the last thirty years have seen more than half of the carbon ever exhaled into the atmosphere—that's since the Industrial Revolution began. China and India have come on to the industrial scene, yes; but we haven't reduced driving our cars, heating our homes, flying off for holidays, eating beef, buying plastic bottles—despite the evidence. In fact, we're increasing our use of fossil fuels. That's why the energy industry, in America and globally, is planning multi-trillion-dollar investments to satisfy the demand. *The Economist* calls it a "Crude

Awakening". ExxonMobil plans to pump 25 per cent more oil and gas in 2025 than in 2017.[40]

Global markets have been given a free rein

For decades now—since the rise of neo-liberalism, and Friedrich Hayek's theory that the market can and will shape society for the good, if allowed to do so—global markets have been given a free rein, such that it's hard for national governments to regulate them. That's even more the case with global companies responding to desires that don't respect national boundaries. As *The Economist* claims, the five oil majors—Chevron, ExxonMobil, Royal Dutch Shell, BP, and Total—have more clout than some small countries. All say they support the Paris Agreement and are investing in renewables, and plan to curb emissions—but that said, they all plan to expand their output. Instead of the recommendation of the IPCC—the Intergovernmental Panel on Climate Change[41]—that oil and gas production needs to fall by 20 per cent by 2030, and by 55 per cent by 2050 to stop the earth's temperature rising by more than 1.5 degrees Centigrade, they all plan to grow the oil industry.

Why? Well—crudely—because we want it. Societies and stock markets still rely on the returns of the oil industry. We can't imagine life without oil and its products. And as Jem Bendell points out, the neo-liberal ideology that has held sway since the 1970s has ensured that we think of ourselves as hyper-individualist consumers, able to change the world through what we choose to buy (light bulbs, sustainable furniture)—and of course, that was never going to be enough, without changing the fundamental economic structures. We need structures that offer more regulation to control the worst excesses of capitalism; economics that serve the common good, with the commons offered by the natural world as the first priority. A key principle of neo-liberalism is market fundamentalism: that the market will correct itself to serve the best interests of society—but that has not proved to be the case. Rather, neo-liberalism has disguised the enormous power of the oil industry to protect and develop its own interests. We have been led to believe that small, incremental steps in the right direction will lead us there—when instead, we've allowed ourselves off the hook, with small congratulatory pats on the back, when we've put this or that sustainability strategy or policy in place. We've adopted atomistic approaches that have meant political and economic change hasn't happened as it needed to, developing a different economics to the dominant neo-liberalism, with the sustainability of the planet as the overriding priority.[42]

It's not too late to rescue things, though. We need to attend to what motivates us, and those around us. In August 2018, Greta Thunberg, then a fifteen-year-old schoolgirl, held a solo protest outside Sweden's parliament. Globally, now, millions of schoolchildren and adults are following her example, striking from school and work, motivated by fear and by frustration at the lack of action by governments.[43]

The next fifteen years are critical for climate change. If innovators, investors, the courts, and corporate self-interest can't curb fossil fuels, then the political will has to be found. That depends on voters: on people becoming motivated—by hope, yes, but also by fear—to change and to demand change. We need to start marching where the lead is—with the schoolchildren of the world. With Extinction Rebellion.

The countryside around us was beautiful, that May and June. This was before it turned into the driest summer on record.

The environment is under threat as never before. Instead of moving towards greater global co-operation on issues that require global solutions—on the environment, on security, on poverty—we are becoming more tribal, with nationalism and regionalism on the increase. The best global leadership in recent years has been offered by Pope Francis in his encyclical *Laudato si'*, which is an excellent piece of work on the global threat of climate change, showing the global reach of the Catholic Church. But otherwise the nations of the world are retreating from international co-operation into silo mentalities which undermine international collaboration.

Hope is not easy optimism

It is a challenging world in which to sustain hope. For it to be real hope, hope that doesn't collapse into false optimism or despairing pessimism, the threats need to be taken seriously. Hope is not easy optimism, but a deep-seated understanding of the way God brings good out of the direst circumstances, life out of death. As a Christian, I need a hope that does not disappoint, and the ability to give an account of it.

It's easy to feel entangled, as in a net, unable to change anything, caught and bound, snarled as a dolphin unintentionally captured in a fishing haul. Dragged, drowning in an invisible membrane, struggling wildly to be free; entangled with cords too strong. Entanglement is this: the silencing of hope.

To enter that death and transform it can be possible if we embrace the metaphor and change it. Like the cloud of unknowing which becomes a mantle of God's grace, we can enter into the entanglement. We can begin to see ourselves, and to grow hope, inspired by the sense that we are all entangled with the environment, with the world around, other peoples and nations, in a world that needs co-ordinated action. Political entanglement, one with another: sustained and energized by the spiritual sense of God's entanglement with creation, in which we participate.

Entanglement is used metaphorically in many different ways. For instance, we could move in a different direction by considering a tiny packet of energy and matter, a very small speck of something. A photon of light, perhaps, or an atom of matter. We know, because physicists tell us, that such particles give us the world around us. Incomprehensibly small, some more elusive than others; present in infinite number and variety.

The connection is maintained

Back in the 1930s, Erwin Schrödinger wrote to Albert Einstein, describing a strange phenomenon in the newly discovered quantum theory. He had noticed that particles, connected, became matter. Yes. But a strange thing happened when you separated the particles again, even to some great distance. For the connection was maintained. However far distant in time or space, each particle mirrored the action of the other exactly and instantaneously. Schrödinger coined the word "entanglement" as he wrote to Einstein.[44] This theory of quantum entanglement suggests that locality is different from the way we understand it in the everyday world. Particles relate to each other across place and time, perhaps a little like invisible threads that resonate in woven fabric.

Physicists today find words inadequate to describe what they discover. The horizons of knowledge leave existing language behind. So we hear of charm and spin, of fermions and bosons, of quarks and leptons: new words to name strange phenomena as humanity tries to describe the fundamental nature of the universe—the seeming paradox of particles which are also waves, energy that is light. Throughout the material world there are phenomena that make no sense to our human minds in a universe that is ultimately mysterious to us. Each particle echoing, resonating with others, in myriad relation. The skylark's song a thread, a wave of sound, that illustrates time and space entangled.

Catherine Keller, in her book *Cloud of the Impossible*, elaborates the planetary ways in which we are entangled, politically, socially, ecologically, and says (as Schrödinger does) that it is only in relation, in non-separability, that we know anything, and find new possibilities of flourishing.[45] She offers fruitful theopoetic reflection on how the word "entanglement" can help us imagine how God relates in and to the world. She reminds us that Schrödinger wrote that "physical action always is *inter*-action; it always is mutual." This is true of how the natural world is observed, for the presence of the observer has an impact on the observed. Perhaps, suggests Keller, this is how God is known:

> There is a disturbance, an impact, what Schroedinger called an "impression," not just from the observed upon the observer—but *reciprocally*: "the observer makes an impression upon that which she is trying to observe. The object I am trying to observe refuses to behave as an object; *it won't stay still*." The object, in other words, no longer permits objectivism. It simply will not make its appearance *outside of its relation* to its observer.[46]

We are entangled one with another, God with the world, human and matter together. There is a participation, where particular beings find what they are meant to be by their participation in the whole, who is God. As *The Lark Ascending* displaced the water which held the boat afloat, we made our impact on the world around and were entangled. We were entangled with the water that was wave and particle, molecules and flow, that moved with us through the locks, often with great force; and at other times was still, mirroring the cloudless sky of that summer. Each particular thing, being what it is meant to be, sustained within the love of God. We were entangled, also, with the people we met, with the environment around, within our lives—past, present, and future. Entangled as a nation, but troubled to its heart. We were entangled with God. Entangled within clouds of unknowing, seeking to know, to see clearly, even as the skies were blue and the sun shone, where there was always more to know and not-know.

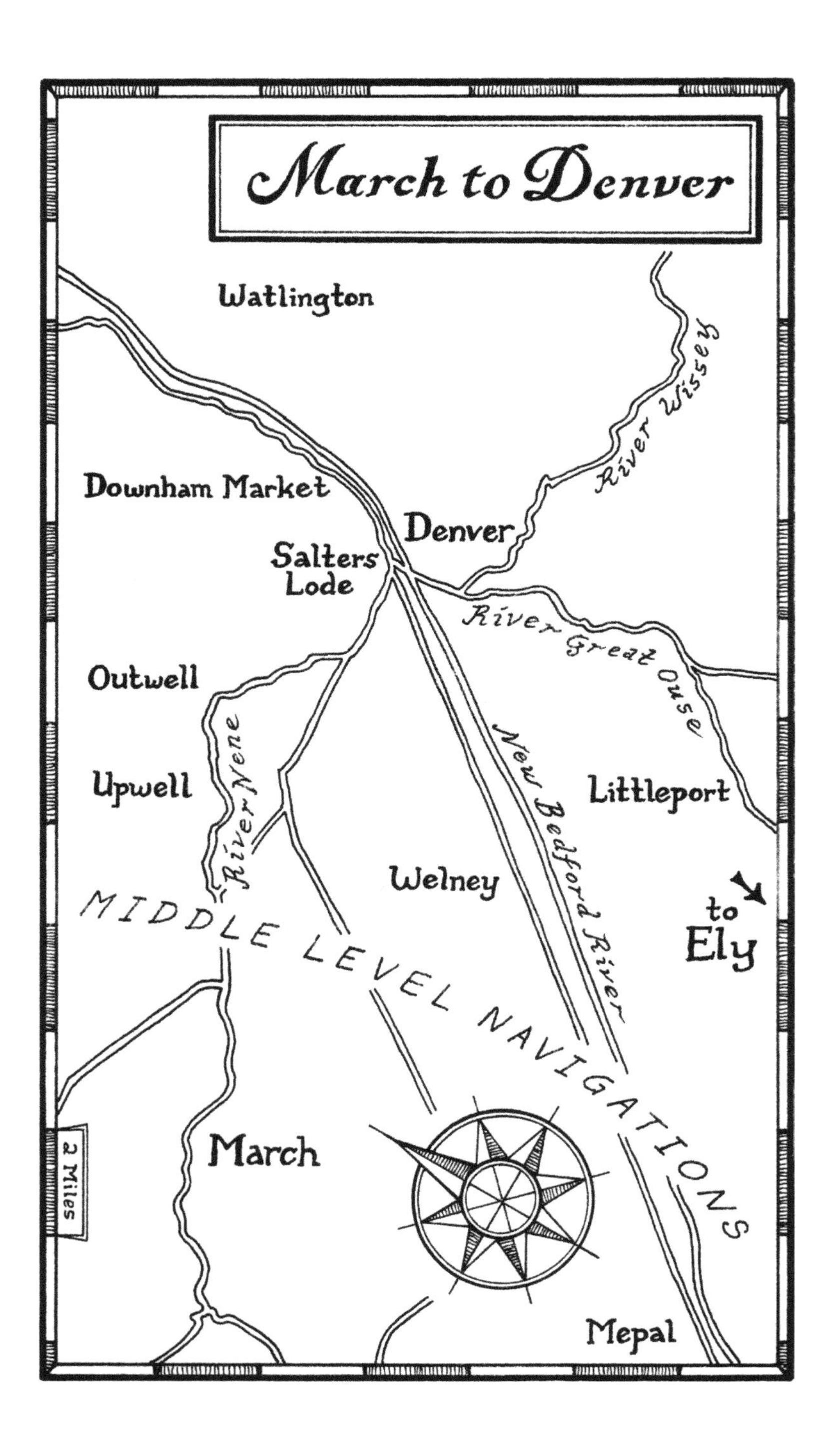
March to Denver
Watlington
River Wissey
Downham Market
Denver
Salters Lode
River Great Ouse
Outwell
River Nene
New Bedford River
Upwell
Littleport
Welney
to Ely
MIDDLE LEVEL NAVIGATIONS
2 Miles
March
Mepal

CHAPTER 3

Larkrise to Skipton

And so it was that my old friend Viv and I came aboard, on Monday 30 April, with a sense of trepidation and excitement at the adventure that lay ahead. We threw ourselves into preparing for our two weeks together, kitting out *The Lark Ascending* and wondering how far we'd get before my uncle's widow, Jenny, joined me, taking over from Viv as she went home on 14 May—wherever we were. My mother's brother had died the previous year in Australia, and Jenny had expressed an interest in seeing England from this different perspective. The plan was to travel the Middle Levels to Stanground, near Peterborough, to travel the length of the River Nene to Northampton, then the Grand Union, the Oxford, Coventry, Trent and Mersey, the Macclesfield, Rochdale, and Bridgewater Canals to Wigan, where the Leeds and Liverpool would take us north-east to Skipton.

"He's painted you a lark as well," said Tracey, Charlie's daughter, as she told us the signwriter was done—the last task before she handed over the boat. "He's written it diagonally. It looks good." And so it did. *The Lark Ascending* was picked out in cream, against a navy blue background. The boat had a cream roof and a green line to edge the sides. The lark was depicted with care, without flamboyance, like the bird itself, in all its insignificance. How can it make such a glorious sound?

Viv and I brought our clothes and other necessities on board, stowing them away in lockers and drawers. Food went in the fridge; pots, pans, plates, glasses, and mugs found their homes—as did Viv's two dogs, Molly and Sid. Molly had lost her leg after a major car accident six months before. She's an old dog now and utterly devoted to Viv, with the characteristic attention and focus of any collie. "She doesn't like boats much," Viv worried. "Bring her along," I said. "If she doesn't settle, Sally can take her home again after the party. She'll want to be with you." No worries about Sid—he's a Dachshund Chihuahua cross, with the temperament of a Border terrier (though friendlier); full of bounce, though with legs too short to manage the steps down into the cabin. "That's probably a good thing," I thought to myself. When you're trying to negotiate a lock, a dog under your feet is the last thing you need.

But that was later. For now, the plan was to head off, not westwards, towards the Nene and Northampton, but to Prickwillow, where my husband, Peter, and I had invited friends to a party to bless *The Lark Ascending*—the official launch.

Why Prickwillow? Well, it's on the River Lark, and it seemed appropriate to begin the voyage to Skipton on a river that shared the boat's name. Not only that, the Lark rises at Bradfield Combust in Suffolk, and then flows through Bury St Edmunds, which had been home until relatively recently. We were heading backwards, into the past, before we could make for the future.

"Larkrise to Skipton!" said one of the ordinands in training at Mirfield, alongside Peter.

◆ ◆ ◆

I tried to find a copy of *Lark Rise to Candleford*, a book I'd enjoyed as a teenager for the way it captured the depths of rural life.[47] Now I wanted to read it again. It offered a baseline for a past world of diverse and abundant English flora and fauna and birdlife, as Flora Thompson wrote in the 1930s, recalling her childhood in the 1880s. She wrote at a time of rural revolution, lamenting the loss of the rural England that has been so much a part of our national imagination. We would pass through today's rural England, now no longer the green and pleasant land it once was. Dieter Helm commends a green and prosperous land which requires a different economic approach to be adopted, reversing the ravages of the post-Second World War agricultural revolution. His twenty-five-year Environment Plan would restore natural capital, including biodiversity, so damaged and depleted by short-term farming practices that have decimated meadows and the natural world by the use of pesticides;[48] so changed in the century since Thompson was a child.

The remembrance and yearning is much more than nostalgia. It begins the desire for restoration.

A strange, melancholic sadness

Robert Macfarlane and Jackie Morris wrote and illustrated *The Lost Words*, dismayed at the losses we've seen:

> Once upon a time, words began to vanish from the language of children. They disappeared so quietly that at first almost no one noticed—fading away like water on a stone. The words were those that children used to name the natural world around them: acorn, adder, bluebell, bramble, conker—gone! Fern, heather, kingfisher, otter, raven, willow, wren . . . all of them gone! The words were becoming lost: no longer vivid in children's voices, no longer alive in their stories.[49]

I travelled with a strange, melancholic sadness at the transient beauty of the season, in the month of May as the hawthorn transformed to glory, fulfilling its desire for summer. I dreamt that England's soul could be as a hawthorn in bloom—a vision of restored beauty and biodiversity, to inspire the restoration of how it once was, and can be again.

I kept a weather eye, each day, on the skies, as I watched for the arrival of the first swift, mindful that it has suffered a 53-per-cent decrease in numbers over the last twenty years. Gone was the biodiversity of the world as it was when I was a child. Mark Cocker's *Our Place* was another book that went with us.[50] He describes the devastating effect on English wildlife of decades of intensive farming. In my adult lifetime the numbers and diversity of our national birdlife have plummeted—the RSPB released the charity single *Let Nature Sing* in April 2019,[51] outlining the losses of our (no longer) common British birds.[52] No longer can we take for granted the larks and linnets, corn buntings, flycatchers, goldcrests, lapwings, and kittiwakes that once filled our soul with song. How unspeakable is a bird-less world.

"Hope is the thing with feathers," wrote Emily Dickinson in one of her poems. Alves likens birds to words:

> Flying birds are unpredictable like the Wind: one does not know where they come from or where they are going. Whenever they arrive they work havoc on the order which has been carefully written on the text.[53]

Like Alves's birds, the real ones around us on river and canal—and the lack of them, for as we kill, trap, and cage birds we kill hope—started to work on my anxiety. The ones I saw kindled a sense of hope which began, gradually, to come alive as we drove the boat, attending and attuned to the environment of river and canal. My senses grew aware of God's grace, like birds, animating the air, the water, and the soil with a deep connectivity, entangling us in the world around with a numinous sense of belonging to the changing countryside.

The wild birds flew through my mind

I found myself regaining connection, learning that I belong within God, with the freedom of the Other who transcends our human strategies, processes, agendas, analysis. The wild birds flew through my mind, bringing another way to be and to belong. Verses from the penitential Psalm 102 settled in my soul:

> I am become like a vulture in the wilderness,
> like an owl that haunts the ruins.
> I keep watch and am become like a sparrow
> solitary upon the housetop.
>
> *Psalm 102:7–8*

As we chugged along, we attended to the countryside and wildlife along the rivers and canals, finding that there is still enough to allow the human self to be immersed in the natural world; to find a numinous sense of participation in the fullness of being, to know and rejoice in the creative love of God in every particle of a beautiful world, even as we lament its desecration and loss.

Turning lament, the deep sadness within, into poetry

I found myself turning lament, the deep sadness within, into poetry, anticipating the change of season that would surely come—when perhaps swallows would leave for the last time. One of my own poems, "Swallow", is a villanelle, which intensifies and transforms the emotion by repetition:

> Our sweet, sweet days are over –
> keened the swallow as she flew –
> by the slow and gracious river.
>
> It is not kind, late summer –
> the days of warmth are few –
> our sweet, sweet days are over.
>
> The cold, cold air and weather
> is not what once we knew –
> by the slow and gracious river.
>
> Crying high above her
> the buzzard's mournful mew –
> our sweet, sweet days are over.
>
> She banks, she turns to hover –
> familiar place to view –
> by the slow and gracious river.
>
> We cannot be together –
> the autumn winds blow true –
> our sweet, sweet days are over
> by the slow and gracious river.

When the wind blows across the Fens

Monday, the last night of April, was the first night we spent sleeping aboard. We were moored up at Salters Lode on Well Creek, waiting to cross the tidal section and to pass through Denver Sluice lock, onto the River Great Ouse.

It was a cold and windy passage, once we had said goodbye to Tracey and Alan at Fox Narrowboats, through the town of March with its riverside gardens and boathouses. In *Green and Prosperous Land*, Dieter Helm speaks of how important gardens are to fostering a diversity of wildlife. Instead of neat, manicured lawns, the product of fertilizers and weed- and moss-killing chemicals, we need untidy piles of logs for hedgehogs, rotting, full of beetles and fungi, and areas of long grasses and flowers. We need to rewild our gardens. The gardens of March did pretty well. Lots of kingfishers, and wilderness along the river through the town—a great byway of abundant life.

Then out onto the Fen of the Middle Levels. When the wind blows across the Fens, not much stands in its way. Houses cower behind the dykes. Humanity crouches, aware of the threat of water that has been forced to retreat. Yet I felt more hopeful, the greater the distance we covered. The real world was around me; it was allowing me in. I found myself looking intently at the banks, into the water, greedy for the finest detail, the smallest strand of the tissue of God's realm. The Fens give you wide horizons, large skies, and this day they were full of clouds, high and fast-moving. In *The Lark Ascending*, we were able to see over the dykes for miles around, over land that once was water.

Graham Swift's *Waterland* is a great novel of these parts, weaving together the history of the drainage of the Fens, the story of the eel, and the tragic account of past events, all of which capture the strangeness of this watery land. Old Cricky, the history teacher, tells the youngsters of his class:

> When you work with water, you have to know and respect it. When you labour to subdue it, you have to understand that one day it may rise up and turn all your labours to nothing. For what is water, which seeks to make all things level, which has

> no taste or colour of its own, but a liquid form of Nothing? And what are the Fens, which so imitate in their levelness the natural disposition of water, but a landscape which, of all landscapes, most approximates to Nothing? Every Fenman secretly concedes this; every Fenman suffers now and then the illusion that the land he walks over is *not there*, is floating . . . And every Fen-child, who is given picture-books to read in which the sun bounces over mountain tops and the road of life winds through heaps of green cushions, and is taught nursery rhymes in which persons go up and down hills, is apt to demand of its elders: Why are the Fens flat? To which my father replied, first letting his face take on a wondering and vexed expression and letting his lips form for a moment the shape of an "O": "Why are the Fens flat? So God has a clear view . . . "[54]

Through this land that floated, as if it were a liquid form of Nothing, we went, until we reached Marmont Priory Lock, taking us from one level of nothingness to the next. We needed the help of the lock keeper there. She lives in a house miles from anywhere.

Marmont Priory Lock

I moored up, still unsure of my skill in bringing *The Lark* alongside the quay, particularly in the crisp breeze that blew. Narrowboats are surprisingly susceptible to wind.

With the dogs in the cabin below, we went to ring a bell for the lock keeper to come and help. She came to turn the handle that raises the paddles of the gate—deep under the surface—letting the water out of the lock so the level was the same as the boat. Viv watched and did as she did, until the water levels allowed the gates to be pushed open. I handled the boat in, with walls dripping with weeds and slime on either side. The gates shut behind me, and Viv helped to raise the paddles on the other gate, so that we rose slowly, slowly until, again, as they had been when we arrived, the water levels in the lock were the same as the river beyond.

Then on we went, having thanked the lock keeper for her trouble (we should have left a tip, but were too preoccupied).

It was cold. I was worried that Viv would want to go home. We were undecided whether to stop. The new wood burner tempted us. But we pushed on, and finally moored up at Salters Lode at about eight o'clock.

Originally there was a Roman Road from Swaffham and the Devil's Dyke to Denver and from there, on older silts, over the Fens we travelled through—Upwell, Outwell, and Nordelph—to March, Whittlesey, Stanground, and on to Peterborough. The Romans started the process of draining this area, but their works fell into disrepair and the land reverted to its natural state of watery bog and fen. Now, as we drove along, we were often many feet above the surrounding land and houses. The peat soil has shrunk or blown away, depleted over the centuries.

The great draining

The story is a long and fascinating one: the draining of the Fens and the forming of the waterways that we travelled, making our way to the river Lark. By the thirteenth and fourteenth centuries, the estuary at Wisbech had become clogged, so the water of the Nene and the Western Ouse flowed increasingly along the Well Creek we travelled. Then navigation was obstructed at Outwell, so in 1478 the Bishop of Ely, John Morton, built Morton's Leam to carry the Nene from Stanground in Peterborough to Guyhirn, thus setting the precedent for all the major seventeenth-century drainage works. In 1605, Sir John Popham, hated by the people for taking their land, cut the straight Popham's Eau north-east of March to the Well Creek. The Fen people objected vehemently to this drainage with its destruction of their way of life of reed-cutting, fishing, and wildfowling. The Fen Tigers were the Luddites of the Fens—they wrecked the sluices and filled in ditches.

A more regional vision was needed, and in 1630 a group of landowners approached Francis, fourth Earl of Bedford. Sir Cornelius Vermuyden was appointed. Between 1631 and 1637, Vermuyden improved many existing drains and built the New Bedford River, which ran from the new sluice at Earith to Salters Lode. By the end of the seventeenth century, the water drained away, the peat dried and contracted, and the land sank lower than ever.

This is Dorothy L. Sayers country

Sayers lived for twenty years as daughter of the vicarage in Bluntisham, near Earith, and set *The Nine Tailors*, her novel of 1934, at Upwell (called Fenchurch St Paul in the book). Lord Peter Wimsey's elder brother is the sixteenth Duke of Denver. In *The Nine Tailors*, the authorities have made a new cut to help the drainage—but it simply increases the flow of the tide on already pressurized sluice gates. The sluice keeper explains to Lord Peter just what the problem is:

> "Nobody knows whose job this here sluice is, seemin'ly. The Fen Drainage Board, now—they say as it did oughter be done by the Wale Conservancy Board. And they say the Fen Drainage Board did oughter see to it. And now they've agreed to refer it, like, to the East Levels Waterways Commission. But they ain't made their report yet." He spat again and was silent.
>
> "But," said Wimsey, "suppose you got a lot of water up this way, would the gates stand it?"[55]

They don't. The sluice goes.

> The whole world was lost now in one vast sheet of water. [Lord Peter] hauled himself to his feet and gazed out from horizon to horizon . . . Away to the east, a faint pencilling marked the course of the Potters Lode Bank, and while he watched it, it seemed to waver and vanish beneath the marching tide. The Wale River had sunk from sight in the spreading of the flood, but far beyond it, a dull streak showed where the land billowed up seaward, and thrust the water back upon the Fenchurches. Inward and westward the waters swelled relentlessly from the breach of Van Leyden's Sluice and stood level with the top of the Thirty Foot Bank. Outward and eastward the gold cock on the weathervane

> stared and strained, fronting the danger, held to his watch by the relentless pressure of the wind from off the Wash . . . The Fen had reclaimed its own.[56]

The Fens are not what they seem

Waking the next morning, already I'm used to the motion which is a constant reminder we are in a different element now; no longer on sure foundations. Here, on water that is feet above the surrounding countryside, it's difficult to tell what is sure and secure, and what is not. The Fens are not what they seem. They make us aware of weather, clouds, wind, and flow in strange ways; of a constant war between water that wants to flow to the sea, and the tides that want to flow over the land:

> The floods have lifted up, O Lord,
> the floods have lifted up their voice;
> the floods lift up their pounding waves.
> Mightier than the thunder of many waters,
> mightier than the breakers of the sea,
> the Lord on high is mightier.
>
> *Psalm 93:4–5*

The floods of 1953, returning in 1998, leave you wondering—as Andrew Hunter-Blair does—just how long it will be before the warning of the Board of Agriculture in 1925 becomes reality: that the Fenland will "return to primeval conditions".[57] The land is farmed extensively—beyond viability, many say—with the soil now exhausted in significant areas. Dieter Helm repeats the oft-quoted suggestion that many soils have only around a hundred harvests left in them. It's not only that, though, for over-intensive farming can release carbon stored in the soil. He writes:

> Part is simply caused by its physical loss, and in some areas of the country, like the Fens, the declines are exacerbated by the dry conditions and wind-blown erosion. Given too that the Fens have large peat deposits, the carbon emissions from soil losses

> are significant, and the costs of reducing these emissions through changing farming practices should be offset against the costs of doing so in the energy sector.[58]

When Viv and I took the dogs walking, we were dismayed at the sterility of the soil; with nothing but potatoes and sugar beet growing. Throughout the river basins that feed the Fenland Rivers, industrial and residential developments continue, with the paved roads feeding water at unsustainable rates into the rivers. Ever more efficient agricultural drainage has the same effect. As climate change means greater rainfall, all this water must find its way to the sea over land which is below sea level, in rivers whose gradients are decreasing and whose floodplains have been reduced. It makes you wonder what the future holds. What are the Fens meant to be?

From Salters Lode we were ready, on Mayday morning, to drive through the lock that stops the tidal water flooding up Well Creek—quite possibly the sluice gate that went, as Lord Peter stayed in the village of Fenchurch St Paul—and out onto that tidal Bedford River. Then we'd take the rising tide a third of a mile upriver to Denver Sluice, and through the lock that lies there, beside the great gates, onto the River Great Ouse.

Denver Sluice

This massive sluice is testament to the human effort to keep the water at bay.

High tide is at 10 o'clock on 1 May, and an hour before, under instruction from lock keeper Paul, we enter the lock and wait for the water to rise on the tidal stretch beyond, to equal the level of the river, but not so high that the boat won't clear the guillotine gate.

Paul talked me through the hazards of this tidal lock as we waited for the water. I asked him if the sandbank was still there, on the approach to the Denver lock. We'd almost grounded on it last year. He said it had been dredged, so to head straight into the lock, or moor at the pontoon. He grumbled about the authorities—just as the sluice keeper does to Lord Peter. He keeps asking them to provide a map of the river. "There's a tree submerged halfway along on the left, so stick to the right. They won't provide a map," he said. "That'll make them liable. Last year I saw a boat that almost went over. It was coming down on the tide and turned, as you have to do, across the flow. It got caught and almost rolled. I thought it had. I thought they were goners. So did they. Screaming in fear they were. The water came up to the windows, imagine that. My whole body reacted with the shock," he went on. Not much shocks Paul. "It'll take a death for them to sit up and do something. And then it'll be to spend tens of thousands getting in consultants who won't have a clue."

He waved as we sharp-turned out of the lock and upstream, and turned back to the lock for the next boat.

The flood-tide water sped us the short distance up the New Bedford River, towards Earith, without hazard. We swung around the sandbank, just in case it was still there, and I brought *The Lark Ascending* alongside the waiting quay, which sits before the massive sluice keeping out the salt from the River Great Ouse and controlling the water throughout the Fenlands. Testament to human ingenuity. How much that ingenuity is needed today in a very different direction. What might it be to let the water in again? As the current advice is that the UK government will need to spend at least a billion on flood defences to protect against rising sea levels, and to relocate communities as flooding becomes a reality, what will happen to these Fens all around us?

CHAPTER 4

Desire Runs Deep

"So you a couple then?"

The inevitable question of Viv and me, as we are both in our late fifties and look like we might be.

It comes as we're side by side with another boat, in the lock at Denver Sluice. It's a deep lock, cavernous, with enormous guillotine gates each end to stop the tide flooding through into the River Great Ouse. The tide's not as high as yesterday's, since the Springs wane with the moon. It feels like an achievement to pull into the lock without mishap, without bumping their boat or worse.

A peroxide shade of grey

Viv was in the bow, alongside the woman on the boat, holding the painter (as the rope's called) around a pole that allows the rope to slide up and down with the water. I was at the tiller, chatting to the bloke, with rope secured in similar fashion.

Viv had asked if she lived on board. He did, she didn't: "but I come for the voyages," she giggled.

Viv explained that no, "That's Frankie; she's married to Peter," and he would be joining us for the weekend on Friday. That she was married to Sally, who lived in Norfolk.

The turn to the personal invited an exchange of information. The woman stage-whispered something, Les Dawson-style.

"Sorry," Viv apologized. "I didn't catch that."

"He's the master; I'm the slave," she repeated to Viv, audibly this time.

We couldn't hear any of this, the "master" and I, a boat's-length away down the lock. Viv covered her surprise and said something non-committal, not sure if she wanted to hear any more, but definitely intrigued.

"That's right. He's the master. He thinks he's in control, but I know better!" She chortled. "We're into BDSM."

The woman was in her early sixties, with peroxide blonde hair pulled back carelessly into a pigtail—a strange shade of grey. She was buxom and jolly. He looked like Jeremy Corbyn.

Later, as we drove away, with them in pursuit (or so it seemed), we googled BDSM. We had got the SM bit but weren't sure what the BD stood for. There's a community out there—or so Wikipedia claims—of folk into all sorts of bondage and domination.

I felt rather innocent. And happy to stay that way.

Each boat a subculture

Narrowboats offer a surprising degree of freedom. You're always on the move so there are no neighbours to gossip, only strangers to tell. And once you're on board, you retreat into a private space, most of it under the waterline. Who knows what goes on in all those boats moored alongside the river and canal banks throughout the land? It's a potent privacy that narrowboats offer. Each boat, a subculture. A private home. Narrowboats enable you to get away from normal routines to indulge the wayfaring instinct that travellers know, to create your own purpose. One steps off the quay, out of normality and onto the water, where things are submerged and flow at different speeds and temperatures, the deeper you go.

God has a clear view over the Fens. So we escape into the depths.

Those depths can unlevel you: stir you out of yourself and normality, to be entangled in other ways. The most obvious, and powerful, is the desire for intimacy so intense you're found and lost.

For it had been, at times, as the psalmist describes:

> Save me, O God,
> for the waters have come up, even to my neck.

> I sink in deep mire where there is no foothold;
> I have come into deep waters and the flood sweeps over me.
>
> *Psalm 69:1–2*

When Augustine talks of the transfiguring desire for God that is never consumed, he knows the power of lust, gluttony, and greed and how they are transformed. The desires that refuse to release us, in the darkest hold, are leavened into something else by God's love. Limbs, such as the body delights to embrace, become part of a divine embrace that is not severed by the fulfilment of desire.

This voyage was the opportunity to attend to the duties, devices, and desires that had held me, but no longer. To acknowledge their power, and to allow God to transform them, me, to a different energy, to the fulfilment of different desires. I found myself on a process of deep transformation. The journey from despair to fierce hope had begun.

Living on water, writing on water, takes us into the subaqueous, rich muddiness and life deep within the stream that flows within us, swirling with stuff we'd often much rather remained submerged.

It's always a temptation to try and see the bottom. William Wordsworth recalls hanging over a slow-moving boat, solacing himself:

> With such discoveries as his eye can make
> Beneath him in the bottom of the deep,
> Sees many beauteous sights—weeds, fishes, flowers,
> Grots, pebbles, roots of trees, and fancies more,
> Yet often is perplexed, and cannot part
> The shadow from the substance, rocks and sky,
> Mountains and clouds, reflected in the depth
> Of the clear flood, from things which there abide
> In their true dwelling; now is crossed by gleam
> Of his own image . . . [59]

He sees weeds, fish, pebbles, and roots of trees, but his view is often confused by the reflection of the sky and clouds, and the mountains around him. He sees, too, his own reflection.

What we see, and don't see

Too often that's all we see, our own image; we are the eternal Narcissus. We remain with what's immediately before our face, the next task, the seemingly urgent and important. There are dominions and powers that want to keep it like that. That's the way our neo-liberal economic system works: we are seduced away into that individualist, market-fundamentalist, incremental, atomistic net of petty concerns which prevent the analysis that might lead to other economic models based on sustainability. We live on the surface and fail to see the deeper life that lies at the bottom. Or, mingled with our own narcissistic reflection—as Wordsworth found, writing of Lake District skies—are the reflections of the clouds. The clouds are there, on the Great Ouse, still and quiet, and soon to evaporate as this hot day dawns. Clouds that draw us away from what we know—our own faces, the task-driven existence—into a right-hemisphere apprehension that there are other possibilities, other ways to live.

Viv takes the dogs for a walk, leaving me alone. I'm thinking about the devices and desires of my heart that have stirred me to my depths, captured my attention. They were beauteous sights; but also dark weeds.

It stirs stuff, living on a narrow boat, surrounded by water:

> I waited patiently for the Lord;
> he inclined to me and heard my cry.
> He brought me out of the roaring pit,
> out of the mire and clay;
> he set my feet upon a rock and made my footing sure.
>
> *Psalm 40:1–2*

Teresa of Avila contrasted the streams she found within her with the clarity of God:

> It is as if we were to look at a very clear stream, in a bed of crystal, reflecting the sun's rays, and then to see a very muddy stream, in an earthy bed and overshadowed by clouds.[60]

The elements mix at the bottom of the river below us—earth becomes water, then water turns into air, to become the clouds, reflected as we gaze. The different layers of observation that Wordsworth and Teresa knew take the observer into the observed, into the water, the mire, and into the clouds too, which are also combinations of water molecules and dust particles. The clouds are very dilute mud, I thought to myself, gazing high in the sky.

Such a darkness and such a cloud

If I'm in the clouds, instead of the mire of desire, as *The Cloud of Unknowing* (written by the fourteenth-century mystic) suggests, I'm taken deep into my own darkness, which is my ignorance of God's grace, driven by fear and despair, those twin sins.

> And ween not, because I call it a darkness or a cloud, that it is any cloud congealed of the vapours that fly in the air, or any darkness such as is in thine house on nights, when thy candle is out. For such a darkness and such a cloud mayest thou imagine with curiosity of wit, for to bear before thine eyes in the lightest day of summer; and also contrariwise in the darkest night of winter thou mayest imagine a clear shining light. Let be such falsehoods; I mean not thus. For when I say darkness, I mean a lack of knowing: as all thing that thou knowest not, or hast forgotten, is dark to thee; for thou seest it not with thy ghostly eye. And for this reason it is called, not a cloud of the air, but a cloud of unknowing; which is betwixt thee and thy God.[61]

My first real thoughts about God were watery ones. Like Catherine of Siena, as she perhaps contemplated Psalm 104, I thought the love of God was best likened to the wideness of the sea, extending beyond the horizon. All we can hope to do is to capture a little of that love to ourselves (I thought then, in my early twenties), as a harbour holds the tidal flow for a little while, and lets it go again. Or rather, we submerge ourselves, little by little, into that great ocean of love.

> There is the sea, spread far and wide,
> and there move creatures beyond number, both small and great.
> There go the ships, and there is that Leviathan
> which you have made to play in the deep.
>
> *Psalm 104:27–8*

And now I read Rubem Alves again. He writes of those who love sailing across the ocean more than arriving at the harbour; of the sea that is not what it seems:

> But there are occasions when the safe, familiar world, comes to its end. Suddenly, the flat land of the ex-plained is interrupted by cliffs and canyons, and it is no longer possible to proceed . . . Or the smooth surface of the ex-plicated begins to crack and one realizes that what was believed to be a solid foundation was nothing more than frozen water, ice which begins to melt, as one's body sinks . . . Sometimes it is because we simply get tired of the safety and boredom of the dry land . . . And we sail to the unknown, obedient to the calling of our soul.[62]

As we have crossed the only stretch of tidal water *The Lark Ascending* is ever likely to know, contemplating the human achievement that has held the sea away over the centuries, I'm left wondering about the overwhelming, inundating flows that can sweep us away. No sluice gate can hold some of the flows of our lives. Perhaps better to let the flow go where it will. And then we must let ourselves go, trusting in God's desire. The only way is to surrender to the cloud and know that my life is in hands other than mine. It is held on or in the great deep, within the dark cloud. This life of mine, with all its silt and mire.

I wrote this sonnet a few years ago, after reading Thomas Traherne:

> And does the sea itself flow in your veins?
> The pounding waves that thunder, suck and draw
> the shore? The surf that licks the sand, then drains
> away, and then, each tide, comes back for more?
> Always the same: a vast similitude

of motion, a heaving, breathing, seething power
that meets, explores, expresses every mood.
Now it reflects the light, some joy-filled hour,
when radiance—jouissance—dances the deep.
Or now the darkness of a grieving sigh
that tunes itself to the distant sob and weep
that pulls forever between the sea and sky.
Traherne thought so. To know the world aright,
the sea flows there, an inner sound and sight.

God's grace breaks through, because God's grace is already there, in sea, cloud, and tree, in stars, and in wild, undomesticated nature. God's grace breaks through, because the world is sacramental. This land knows God and responds with *jouissance*, with rapture, with vibrant dance of atoms. And also, perhaps, with flood and water that cannot be kept at bay for ever. Our desires are there, in the great maelstrom.

The human longing for God

Sarah Coakley uses the Letter to the Romans as one of the bedrocks of her book *God, Sexuality, and the Self*. She writes this about desire, ourselves, and God:

> Desire ... is the constellating category of selfhood, the ineradicable root of the human longing for God.[63]

Our desires find their home, ultimately, in God. Like the stars of a constellation, they find their rightful place as they make us the people we are, as they find their true end in God, in our human longing to be at home in God.

To be troubled by fears and despair that overwhelm you is all-consuming—it covers like a cloud, obscuring the stars. Yet you know that God's presence is there in the experience somehow. The depths are troubled in God, the Holy Spirit moves over the deep; the fear and trembling are known, even when there seems no path into the clear, still water. God's way is in the sea, God's paths in the great waters; but can the right way be discerned? Perhaps only by the stars.

Robert Macfarlane was sailing off the North of Scotland, as he describes in *The Old Ways*:

> My watch turn was at midnight. I groped along to the stern, over sleeping bodies. A whisper from the dark, Diyanne handing the tiller over to me: "It's simpler to steer by the North Star than by the compass. Look for Polaris; it's easy to locate, thought it's not the brightest star in the sky. Find the two stars on the far side of the Plough—see them? Now, follow to where they're pointing, keep going, one star, two stars, and there's Polaris, blazing away. Just hold the North Star steady between halyard and spar and sail on up."[64]

The clouds had obscured my sight; my compass was spinning. I needed to find Polaris again.

> Then would the waters have overwhelmed us
> and the torrent gone over our soul;
> over our soul would have swept the raging waters.
> Our soul has escaped as a bird from the snare of the fowler;
> the snare is broken and we are delivered.
>
> *Psalm 124:4, 6*

All desires known

I've known, with the psalmist, what it feels like to be submerged; to know deep yearning, troubled and unravelled by desires that have taken hold; how desire and despair are bedfellows. I've known what it is to relinquish the pain and turmoil into the hands of God, to escape like a bird from the snare of the fowler. He is dead now. But like the story of Gabriel García Márquez that Rubem Alves retells, he comes back to life as he is remembered, as a dead man washed up on the beach, and washed by the women of the village who did not know him:

> "Did [these hands] know how to caress and to embrace a woman's body?" And they all laughed, and were surprised as they realized that the funeral had become resurrection: a movement in their flesh, dreams, long believed to be dead, returning, ashes becoming fire, forbidden desires emerging to the surface of their skins, their bodies alive again.[65]

I was caught by desires that felt like drowning; and never far away from thought were images of dead bodies on beaches: refugees, seeking new lives, stirring despair.

These homes on water, that are always on the move. These places of intimacy, of living on top of each other. These boats that are here today, gone tomorrow. They stir and bear desire. Never mind that it's canals these boats travel. Never mind that it isn't the sea. The water, that great leveller, holds the flow of desire just as it holds the flow from one lock to the next, the flow of river to the sea. The boat holds life, desire, millimetres away from the deep waters of death.

> Save me, O God,
> for the waters have come up, even to my neck.
> I sink in deep mire where there is no foothold;
> I have come into deep waters and the flood sweeps over me.
>
> *Psalm 69:1–2*

To God, all desires are known. My deepest desires were fluid and changing, aligning with a different flow, my despair baptized into fiercer hope. Early days, for I needed to go deep within me, into the depths of God. To wait for what might emerge. For what a fierce hope might be. How it might become action and change in my life.

You desire truth deep within me

As I spent days and weeks on the boat, the fact of the water beneath me became more and more significant. The foundations of my faith were shifting. Water, the element of chaos from which we emerge, brooded within me, below me.

As the voyage began, Psalm 51 was churning through my heart and mind—one of the seven penitential psalms used during the liturgy on Maundy Thursday. Traditionally called the *Miserere*, from the Latin for its opening words, it's best heard sung; and as I close my eyes, I can hear the St Edmundsbury cathedral choir sing the seventeenth-century setting by composer Gregorio Allegri:

> Have mercy on me, O God, in your great goodness;
> according to the abundance of your compassion
> blot out my offences.
> Wash me thoroughly from my wickedness
> and cleanse me from my sin.
> For I acknowledge my faults
> and my sin is ever before me.
>
> *Psalm 51:1–3*

When you're full of anxiety, weighed down by fear that destroys hope, where else to turn? One of the gifts of faith in God is the knowledge that I belong in God, beyond myself, which enables me to break out of the spiral of self-absorption, the negativity that encircles me. God is the source of goodness and truth, beyond my own creations and figments of imagination, beyond the dominance of my emotion and feeling states.

But when that God is angry? Unforgiving? I prayed, mindfully, with the mantra *Christe eleison*, "Christ, have mercy." Repeated again and again, it forms a cloak around me; I become hidden in God's mercy.

The psalmist sings to God:

> Behold, you desire truth deep within me
> and shall make me understand wisdom
> in the depths of my heart.
>
> *Psalm 51:7*

It is a prayer for a new way of being, a prayer for truth and wisdom that breaks through the self-deceptions that go around and strips me back to the real person I am, in truth before God.

> Purge me with hyssop and I shall be clean;
> wash me and I shall be whiter than snow.
>
> *Psalm 51:8*

I made the words my own:

> Make me a clean heart, O God,
> and renew a right spirit within me.
>
> *Psalm 51:11*

That new and right spirit depends on the presence of God:

> Cast me not away from your presence
> and take not your holy spirit from me.
>
> *Psalm 51:12*

Morning and evening prayer begin with words from this psalm:

> O Lord, open my lips
> and my mouth shall proclaim your praise.
>
> *Psalm 51:16*

It's a reminder to me that my true humanity begins as I turn to God in prayer and praise, relinquishing my sense of self-importance—and my anxieties and the weights on my mind—in the recognition that God is the ultimate other in my life who shapes me in love as I turn again and know myself, as I participate in the fullness of God.

A broken and contrite heart

The psalm calls us to acknowledge that we are not whole. It is a difficult verse, the one that challenges me to face into my own brokenness and to know that I am held.

> The sacrifice of God is a broken spirit;
> a broken and contrite heart, O God, you will not despise.
>
> *Psalm 51:18*

It can be taken to imply that God wants us to be broken and in pieces before we are acceptable to God. The God of love doesn't want this. Instead, it's about being open to God, knowing a lack of wholeness and allowing God's love to break through the cracks in our sense of self that we construct so carefully, to transform us, and remind us that we live and move and have our being in God. Then again, sometimes we need to break. We need to hit the bottom, to burn out.

The poetry of the psalms uses strong, unusual language to make an impact on our souls. This is not careful wording, designed to avoid offence. It's language that you need to swallow and stomach, and digest in your bowels. It takes you to your own anxiety, worry, desolation, anger, and aggression—so often felt in our guts. It allows you, within the structure of the verse, and within the structure of the liturgy in which it belongs, to lose control within the enormity of the otherness of God, in order to find a deeper self-control.

Then, another psalm, and I'm face to face with God's wrath at what humankind does and my implication in the sin that destroys:

> The waters saw you, O God;
> the waters saw you and were afraid;
> the depths also were troubled.
> The clouds poured out water, the skies thundered;
> your arrows flashed on every side;
> The voice of your thunder was in the whirlwind;
> your lightnings lit up the ground;
> the earth trembled and shook.

> Your way was in the sea, and your paths in the great waters,
> but your footsteps were not known.
>
> *Psalm 77:16–19*

It's hard to find peace of mind today. The febrile culture in which we live feels as if it's lost the plot. I feared I had too—no, I knew I had. I had to live by a deeper flow.

Slow with alternative possibilities

The regenerative energy of the English countryside in May speaks of new possibilities. The world from the water, along rivers and canals that once were derelict but are now corridors of wildlife, began to inspire new hope. I was slow with alternative possibilities. I found myself re-engaging with green life, with hopefulness, entering the cloud of unknowing, emerging again and again, baptized, regenerated by God's grace.

John Rodwell, in a series of Holy Week talks he gave at the College of the Resurrection, Mirfield, in 2018, said we need to hone ourselves to hope—and these words are taken from my lecture notes:

> We have to hone our human intellect; shape our moral sense, stretch our imagination; fashion our faith and hope. Nature does not hope, but groans in travail. Nature is the place where we learn how to be human; when we hope we are met by God and discover something of his desires and yearning for us and for creation. We access God's own desire. To go on hoping isn't deluded; just because our own hopes are not answered doesn't mean prayer doesn't work. God's object is to place us in the tissue of his kingdom.

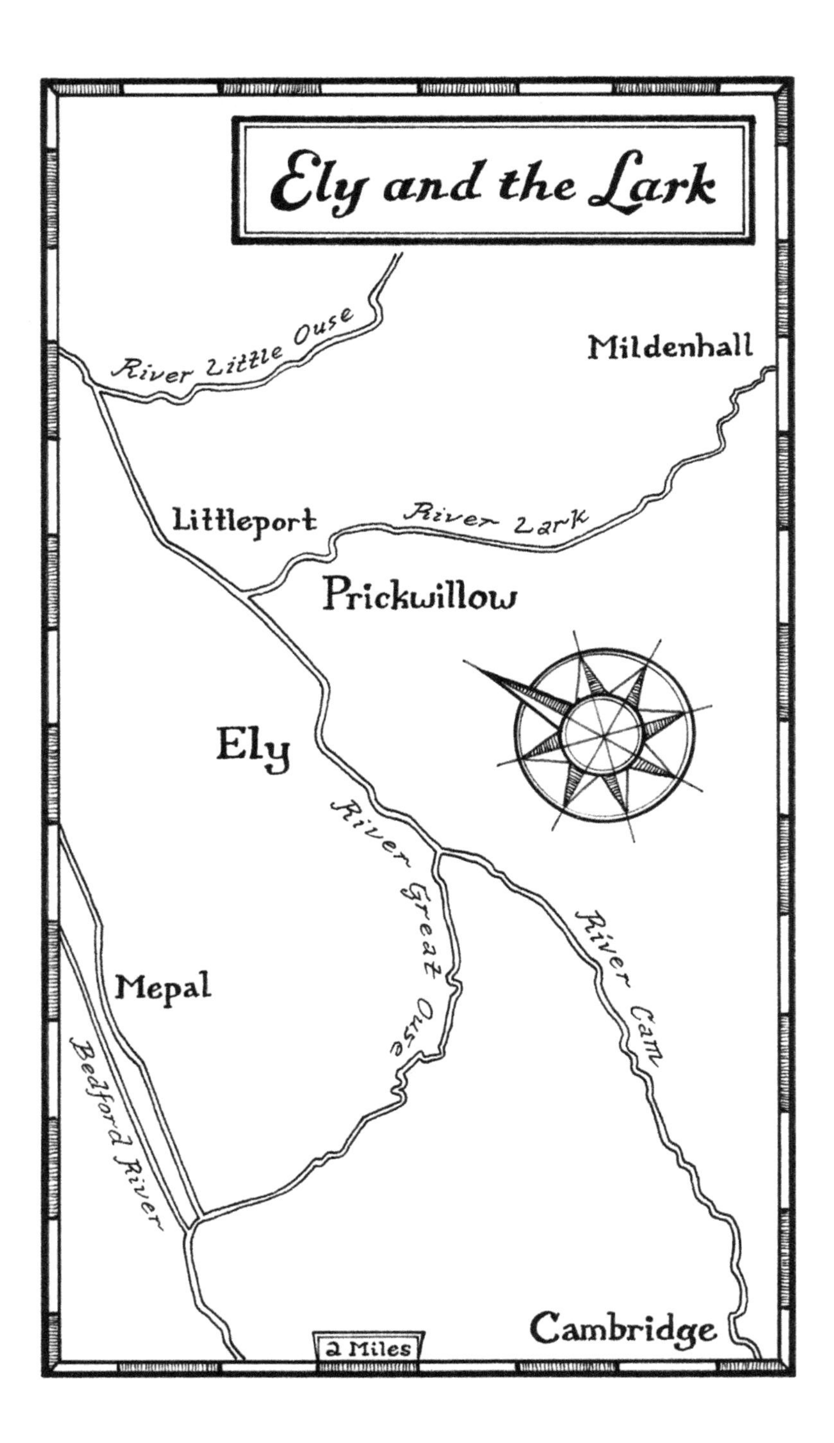
Ely and the Lark
River Little Ouse
Mildenhall
Littleport
River Lark
Prickwillow
Ely
River Great Ouse
River Cam
Mepal
Bedford River
Cambridge
2 Miles

CHAPTER 5

Ely and Prickwillow

Ely is so deeply familiar; I am in its tissue. I know its faces, streets, and trees; the stones shout aloud.

This visit was a goodbye in many ways. When I moved to Bury St Edmunds in 2010, it was a return to East Anglia after time in St Andrews, Cambridge, London, Manchester, and Bradford. This summer Peter and I headed north again, away from more than Suffolk.

Will we ever know "why"?

I spent the 1970s in Ely. The town thrives today; no longer is it the grumpy little place I remember, clustered around and yet repelled by the great cathedral that dwarfs all else. The market is wonderful, the riverside a great place to moor, the antiques warehouse a treasure trove. This weekend, the early May bank holiday, there's an Eel Festival (which we'll miss, unfortunately), with its parade of eels (acrylic) through the town, and other eely games. My brother used to fish for eels when they were plentiful. Now the eel is on the list of critically endangered species, and the Environment Agency is working hard to increase their numbers—particularly of young eels, or elvers, that travel from Bermuda to freshwater rivers in Europe before they return to breed in the western Atlantic. It's hard work, though. There's a raging market for eels and poaching is rife around the world.[66]

Graham Swift, in his evocative book *Waterland*, writes of the eternal question, "Why?":

> Curiosity will never be content. Even today, when we know so much, curiosity has not unravelled the riddle of the birth and sex life of the eel. Perhaps these are things, like many others, destined never to be learnt before the world comes to its end. Or perhaps—but here I speculate, here my own curiosity leads me by the nose—the world is so arranged that when all things are learned, when curiosity is exhausted (so, long live curiosity), that is when the world shall have come to its end.
>
> But even if we learnt how, and what and where and when, will we ever know why? Whywhy?
>
> A question which never baulked an eel.[67]

At this time of the year, everything is blooming, as now the sun shines—all is green, green and full of blossom. The park looks wonderful, full of stately old trees, many of which I climbed as a child, including the old London plane tree (always "the wonky tree" in my mind), which still stands in the Dean's Meadow, propped up here and there, like an old man.

When I was little, I climbed trees

It gives a different perspective on life, climbing trees. One of my favourite haunts as a nine-year-old was at the top—the very top—of the ancient but still fruitful Beauty of Bath apple tree that was in the front garden. I would sit there for hours—surrounded by early blossom in the spring, or better, able to eat, at my leisure, the ripe fruit in early autumn. Supported by familiar strength even when the wind blew. The branches knew their place, so I could ascend and descend with my eyes shut. I'd sit there, held by a living organism, and I'd imagine all sorts of things. Tall ships and stars to sail by. I was mistress of all that I could see; gardens and countryside around—far-reaching perspectives that expanded my horizons and grew my soul. Apple trees; orchards: a great way of capturing carbon; of providing blossom and fruit.

Or later, an old, evergreen turkey oak in the close in Ely where I'd go at night alone, swinging up into the branches and sitting still and quiet, surrounded by the velvet darkness. There and then began the first stirrings of a contemplative inner life that, even now, continues to seek God's presence and to be immersed in silence and in the natural world.

I'm not alone.

Robert Macfarlane walks out from Cambridge, over the fields, to his favourite tree, in search of the wild:

> I came to my tree—a tall grey-barked beech, whose branches flare out in such a way that it is easy to climb . . . . At about ten feet a branch crooks sharply back on itself; above that, the letter "H", scored with a knife into the trunk years before, has ballooned with the growth of the tree . . . Thirty feet up, near the summit of the beech, where the bark is smoother and silver, I reached what I had come to call the observatory: a forked lateral branch set just below a curve in the trunk. I could stay comfortable there. If I remained still for a few minutes, people out walking would sometimes pass underneath without noticing me. People don't generally expect to see men in trees.[68]

And another Robert wrote of swinging on birch trees: Robert Frost, describing how birch trees become flexible with the weight of a boy, and allow one to ascend toward heaven, then gently to return to earth, the right place for love: "good both going and coming back".[69]

Sight became insight

To climb a tree is to be absorbed into nature. It's to be at one with the countryside, full of flowers, birdlife, trees, creatures, light and shadows. Countryside that has inspired poets and writers through the ages, who have so often captured a sense of immersion in the natural world, with delight and the sense of God's presence in the variety and diversity of wayside flowers, each little bird that sings. This is a natural world under threat, a world which, as God's gift to us, requires our attention like never before. As I write, the rainforest burns—all those ancient, magnificent trees and the life they sustain. Everything within me rebels against the news. I must find a way to live with this grief. I don't want this agonizing lament to be normal. I don't want to cry aloud the question, "Whywhy?", but relish and rejoice in the abundance of life. This sacrilege is evil, monstrous, so painful. We are destroying all that life, alive in God's grace, even as it dies.

When I was little, I spent hours outside: yes, climbing trees, and roaming the countryside, my senses alive to sound and smell, to sight and taste. I would make myself invisible, hide myself in the natural world, and in my hiddenness, feel at one with the vibrancy around. I'd come to have a special sort of insight, if I looked really closely at the particular flower, or stone, or bird before me. My sight became insight. Then the light, the silence, the wind, the colour around me came to have an extraordinary vitality, a different reality. It was wonderful; it left me full of wonder. The transcendent was one with the immanent; the beyond in the here and now. It was a mystical sense.

◆ ◆ ◆

I walk along The Close in Ely, passing the gate to our home that was, with its long garden, squeezed between the Bishop's House and the Priory. We had hedgehogs galore back then, and Mum was always digging up signs of monastic living from centuries ago—oyster shells, fragments of this and that. The old pear tree is still espaliered against the top wall, where one spring I put a milk bottle over a growing pear until there it was, trapped inside, freedom only possible when it rotted, or the bottle broke. And there was the drawing room window, through which I'd blasted The Rolling Stones' *Rolled Gold*, recently released. The then Bishop, with his study window at right angles on the same first-floor level, never complained. Perhaps he liked it.

The Cathedral is now resplendent with a lovely new altar and furniture—thanks to a legacy from the Rt Revd Peter and Mrs Jean Walker. He was Bishop of Ely from 1977 to 1989. I remember his wisdom as I began to explore my own vocation to the priesthood. How he talked of the priest, the preacher, the prophet, and the pastor, how each needs to be owned and developed as a gift. He might also have talked of the poet, with his love of W. H. Auden.

The altar is a great memorial. It is octagonal and seems to have been lowered from the Octagon above; a lovely rich colour and shape that holds the space perfectly, without drawing attention unnecessarily to itself. Elegant, contemporary, and fitting. The old choir stalls (which were new and radical in the seventies) are gone to Halifax Minster.

I join the Dean and Residentiary Canons for Morning Prayer and the Mass. The office is said reverently and seriously, with silence and care, with a real sense of this being the heart of the corporate life of the Cathedral. The girls' choir, singing on Wednesday, sounded pure and strong. Traditions of prayer and praise that are strong enough to hear my lament and answer it.

Ely takes to ghosts

Ely, with its Cathedral Close, has so many memories. Not all of them happy—it wasn't always easy to grow up there. Every time I return, I lay one or two more ghosts. They are in good company, I think to myself, as during Morning Prayer I contemplate the final resting place of the *ossa* of the seven tenth- and eleventh-century bishops and martyrs, some "*caesus a Dani*" (killed by Danes) in Bishop West's Chapel. Ely takes to ghosts; it knows what to do with them.

After prayers, I walk the familiar way that each Tuesday morning I went, like a snail, to my flute lesson, to visit an old and dear friend. I had the privilege of taking the funeral last September of her late husband, who had valiantly persevered each week, when I was young and struggling with the flute, as I laboured away, always guilty at my lack of practice. He made lutes—and was ordained later in life. We read George Herbert's "Aaron's Drest" at the funeral. Their home was a haven at times during my teenage years—the alternative home everyone needs at times.

◆ ◆ ◆

Thursday morning found Viv and me at the market in Ely, finding all sorts of useful and edible things, and then, on the way back to the river, yet another visit to Cutlacks, who had seen a lot of us that week—a great family business that has been going for decades. We've kitted the boat out fully now, and shared the satisfaction of little jobs done that leave you feeling slightly more sorted than before.

Scholar Gypsy passed on her way along the river. Simon, the owner, was off to Little Thetford to do some writing, but will join us at Prickwillow on Saturday. Good to put a face to the father of Jez, a friend of my son Jonty. I must read Matthew Arnold sometime. I'm looking forward to seeing over *Scholar Gypsy*, curious to see how Simon has arranged the internal space.[70]

And back on the boat, we see King's Ely girls and boys out in sculls, pairs, and fours, and canoes, coached by someone who was on the staff when I was a pupil in the 1970s. Some things really don't change. The boathouses have, though. Cambridge University have theirs, further down the river, sporting their sponsorship from a large American bank. There were those who were incensed at the environmental damage caused as it was built.

Kingfishers, crested grebes, swans, swallows, geese and duck with young, heron—the bird life is still there in some abundance. It was good to see a great crested grebe catch and eat a fish just there, a few yards away from the boat.

Topping & Co wasn't there in the 1970s. It's a great bookshop, bucking the trend of Internet shopping, secure in its reputation for books and coffee. Viv and I sat amongst the art books and poetry, and I showed her my purchases: Jenny Uglow's biography of Edward Lear—a signed first edition![71] Uglow's book on Bewick, the bird engraver, was brilliant, so I can't wait to read this; Alexandra Harris's *Weatherland*—which seems appropriate, with weather so important to us just now;[72] Mark Cocker's *Our Place: Can We Save Britain's Wildlife Before it is Too Late?* (I passionately hope so); and more Iris Murdoch, as I read all her novels. Already ticked off are *The Bell*; *The Sea, The Sea*; *Nuns and Soldiers*; *The Italian Girl*; and now, just finished, her strange last novel, *Jackson's Dilemma*. Written when her dementia had begun to set in, as John Bayley records, it seems as if she is already losing the plot.[73] It is held, but only just—and does Jackson represent her growing enigma to herself? We never get to the bottom of him, never quite understand his dilemma. I'm not sorry to finish it; though perhaps my reading was clouded by the knowledge of her dementia. What if I had read it unaware? What did I miss because of my pre-judgement? Novels take you in; or we hold back. I held back on this one, not quite trusting Murdoch to be Murdoch. Do I do that with people? With myself? Perhaps inevitably.

My God, my rock

I remember, when I was about seventeen, falling into a deep, dark place. I suspect I'd drunk too much coffee (nothing more), enough to send me spinning down into an abyss. It was the beginning of faith, to know that this abyss was not bottomless—that there was a rock below me, on which I could rest and stop that terrible fall. It's there in Psalm 18:

> I love you, O Lord my strength.
> The Lord is my crag, my fortress and my deliverer,
> My God, my rock in whom I take refuge,
> my shield, the horn of my salvation and my stronghold.
> I cried to the Lord in my anguish
> and I was saved from my enemies.
>
> *Psalm 18:1–3*

It was, perhaps, the beginning of the depression and anxiety that every so often take hold. And the sense that I am held, none the less:

> He reached down from on high and took me;
> he drew me out of the mighty waters.
> He delivered me from my strong enemy,
> from foes that were too mighty for me.
> They came upon me in the day of my trouble;
> but the Lord was my upholder.
> He brought me out into a place of liberty;
> he rescued me because he delighted in me.
>
> *Psalm 18:17–20*

It's Psalm 69 that captures it best—that sense of being overwhelmed, out of one's depth:

> Save me, O God,
> for the waters have come up, even to my neck.
> I sink in deep mire where there is no foothold;
> I have come into deep waters and the flood sweeps over me.
> I have grown weary with crying; my throat is raw;
> my eyes have failed from looking so long for my God.
>
> *Psalm 69:1–3*

Addiction does it too. Losing oneself in the depths of drink, for instance, losing control. Reaching the bottom is a blessing then. You can only head upwards. Or learn to breathe under water, as Richard Rohr commends. He commends it, not only to individuals who are overwhelmed by the demons of addiction, but also to institutions that ape the addictive habits of our culture. The first lesson that is learned, if you attend an Alcoholics Anonymous meeting, is your own humility and powerlessness. Rohr writes:

> Christians are usually sincere and well-intentioned people until you get to any real issues of ego, control, power, money, pleasure, and security. Then they tend to be pretty much like everybody else.[74]

Yes, that is me. I am addicted to all the usual things, immersed in the culture that swamps us—good and bad. I long, though, for redemption, and for the world around to be a more sane, lovely, and blessed place to be, where all God's children shall be free.

Goodbye to Ely. It's a place I'll always visit with strange and deep layers of knowledge. I lived there during formative years. Such knowledge cannot be unknown. I know myself as I was then (or think I do), and as I am now, and I grow in the encounter with the past.

Party in Prickwillow

Peter thought it would be a great idea to start the great voyage on the River Lark. It flows through Bury St Edmunds, for a start, and out into the River Great Ouse beyond Prickwillow; the boat's name, another reason. So the idea of a gathering to bless the boat on her way grew, and people were invited, including the Rt Revd Tim Stevens, erstwhile Bishop of Leicester, who agreed to provide an episcopal blessing.

Saturday 5 May dawned hot and sunny, and the preparations began, with the excellent help of family and friends.

Simon—who owns *Scholar Gypsy*—arrived, having walked from Ely along the old causeway that follows the course of the River Great Ouse, straightened in 1829. Before that the Ouse flowed to Prickwillow from Ely, but then it was diverted, and the old course was filled in and ploughed. The draining of the Fens continues its impact—in the 1920s the constant draining of the land meant it was sinking by two inches a year. St Peter's Church was built in 1866, on piles to ensure stable foundations. The Vicarage had two steps put in to reach the front door, and as the land eroded away, more steps were added. The original cellar is now the ground floor. People can't be buried here any more, as the water table is so high, so they end up in Ely. The local road to Isleham, the B1104, is said to be the most affected by subsidence in the country. The six-and-a-half-mile drive causes seasickness—the water's revenge, perhaps.

Gradually we gathered—by road, by foot, by boat.

Tracey and Gary came from Fox Narrowboats in March. And so many East Anglian friends—from old days at King's School Ely, to friends from St Edmundsbury Cathedral, Cambridge, Ipswich. Then the blessing of cello and violin emerged from *Scholar Gypsy*, and Vaughan Williams sounded out across the water—a memorable performance of *The Lark Ascending*. Bishop Tim was suitably and ceremoniously attired in mitre and cope, and his words captured the occasion perfectly, with a blessing conveyed in word and river water splashed liberally on Peter and me, and the boat, inside and out.

A joyride downstream for those who wanted, with drama from Jonty who left *The Lark Ascending* by suspending himself from the rail bridge to wait for *Scholar Gypsy*, coming along behind.

It's an interesting river: like so much of the water and land we passed through, or along, not what it seems.

One of only 200 chalk streams in the world

When living in Bury St Edmunds, I used to jog along the banks most mornings. There, you'd hardly call it a river as it rarely rises to a flow. Between it and its sister river, the Linnet, is an area called the "Crankles" (originally ancient fish ponds) and No Man's Land. This was where the monks had their allotments, and you can still see the furrows of their agriculture. The rivers are chalk streams, dating from 8,000 years ago, and they meet just by the East End of the old ruins of the monastery—a prime spot for the monks to build. The River Lark rises to the South of Bury, near Bradfield Combust—the Linnet in Ickworth Park, south west of the town. They are not as big as they were in the Middle Ages, when they were used for transport. It's said the stone for the Abbey came by river. No Man's Land and the Crankles still function as water meadows—I've seen them flooded on occasion in winter. But that's unusual. The water table is low now, with extractions from local agriculture and businesses. The Bury Water Meadows Group are doing what they can to care for and develop the potential for wildlife and the environment.

So *The Lark Ascending* can't get as far as Bury St Edmunds—nothing like. We could have reached Isleham, where there's a marina, and then beyond to Judes Ferry where there's a winding hole. Instead, I'm going to turn to a journalist local to Bury, a friend of a friend, Matt Gaw, who canoed as much as he could of the Lark. He tells the story of the river, from Bury St Edmunds to Prickwillow.[75]

It was one winter, recently, that Matt set off with a friend to canoe the river. Or to try to. He writes in *The Pull of the River*:

> The Lark. The clue is in the name. A place that should sing and burble in flight. One of only two hundred chalk streams in the world, it ought to be revered and loved, not subdued with concrete, broken by sluices and cruel flood defences. It is a source of shame, or bloody well ought to be.[76]

Matt tried his luck from Tesco in Bury, hoping to float through the Brecks and the Fens, past Prickwillow, and to the Great Ouse. "A few inches of murky water dribbles over a slimy lip down to muck-covered stones. Looking upstream, the river really is boxed in: separated from its floodplain and starved of any of the natural processes that could elevate it above a giant car-park puddle."[77]

He and his friend have to walk, it's so shallow. The litter is bad, really bad. Matt points out where the coal road crossed the river; where St Saviour's Wharf used to be, when the river was once busy with trade. Down towards Fornham, where there's more rubbish to see. The water flow still isn't great—he describes it as a "winterbourne", flowing only from source to mouth from late autumn to early summer. Dry in the summer; the banks invaded by exotic species, including the American signal crayfish. They saw cages to capture the mink—so destructive of local life. The lumbering coypu have gone. All are "the wrong kind of wild," as Matt remarks.[78]

By the time he gets to the village Fornham the river has opened up: "the water is low, but deep enough for us to canoe, and it riffles and skids over the bed, a sucking, chuckling burble. It feels like the Lark is singing again."[79] The Bury St Edmunds Trout Club have worked to transform the Lark from a straight, dredged channel into a sinuous, bustling river. It's a success as a rewilding project: "trout are already beginning to return and so too has the sound, the slowing and speeding water trilling and singing like the most beautiful bird. The Lark has found its voice."[80]

They resume their voyage at Judes Ferry—for there's no way over a sluice, and the dual carriageway of the A14 is too busy to cross with a heavy canoe. It's beautiful from Judes Ferry—I know. Peter and I have walked the banks. The water is deep and clear. Matt reminds me of the ancient custom of baptism here. I knew that the great nineteenth-century Baptist preacher Charles Haddon Spurgeon (1834–1892), who founded one of the leading Baptist Colleges, was baptized in these waters. In 1850, after being converted away from his nominal Anglican background, and while staying in Newmarket, he made his way to the River Lark at Isleham where he was baptized on 3 May (168 years, almost to the day, earlier than our party). He moved to Cambridge, where he became a Sunday school teacher. He preached his first sermon that winter in Teversham, in a cottage, and that was the beginning of his public ministry, which continued as he was installed as pastor of the small Baptist church at Waterbeach, near Cambridge. He was quickly recognized as someone with extraordinary gifts.

Cradled in all the oceans of time and space

Matt Gaw found the story from Spurgeon's autobiography:

> The wind blew down the river with a cutting blast, as my turn came to wade into the flood; but after I had walked a few steps, and noted the people on the ferry-boat, and in boats, and on either shore, I felt as if Heaven, and earth, and hell, might all gaze upon me; for I was not ashamed, there and then, to own myself a follower of the Lamb. My timidity was washed away; it floated down the river into the sea, and must have been devoured by the fishes, for I have never felt anything of the kind since . . . I lost a thousand fears in that River Lark.[81]

It is a lovely stretch, downriver of Isleham. To swim there, as Peter and I did the summer before from one of Fox's hired narrowboats, is to know its beauty: so quickly out of my depth, a few strokes took me a good distance from the boat—then to lie there, on my back, with eyes open to the skies, allowing the water to carry me, buoyed up by the heavy depths below. The water plays with your body, rippling over and through you, nudging and reminding you that we are largely water ourselves. I breathe in, my chest fills, and I rise noticeably; then out, and I sink slightly, back towards the depths. My ears are full of water; I have entered another element. I trust it so, I could fall asleep, cradled in all the oceans of time and space. Losing a thousand fears.

Baptism does this. It begins a life of immersion in the grace of God, receiving a tradition that holds and shapes one as the water supports life. The Lark is entirely human-made today. It would have been, originally, meandering courses of brackish water, bog, meres, and reed beds, breathing deeply as the tides ebbed and flowed. Now it flows as straight as you like, to Prickwillow, and beyond to join the Great Ouse. The water is clean now. Refreshing the soul. Turning any number of fears to hope.

◆ ◆ ◆

When all the guests had gone, Peter headed off for Mirfield, and both boats paced down the Lark to join the Ouse where we bore right, and off towards Denver. At seven o'clock we moored up against the bank, and watched the sun set over the Fens as we tucked into a barbecue on the bank.

Chocolates and songs on board ended a perfect day.

Leaving the Lark began our forward journey—Larkrise to Skipton.

CHAPTER 6

A Lament for Creation

The sacramental and analogical approach of theopoetics helped me give words to the yearning and lament that I felt.

> What would the world be, once bereft
> Of wet and of wildness? Let them be left,
> O let them be left, wildness and wet;
> Long live the weeds and the wilderness yet.[82]

Deep in the shadow of anxiety about the future of planet Earth, I had to find a way to drive a taproot deep into the ancient wisdom of which Ellen F. Davis writes, if only to find a way to live with the worry:

> The biblical wisdom literature is of great value to Christians as we seek to deepen our theological understanding of creation and thus respond *out of our faith* to the ecological crisis that currently threatens its well-being. One of the pronounced features of this literature is its focus on creation as the realm where God's will and action is to be discerned . . .
>
> "The Lord *by wisdom* founded the earth": This brief statement sets forth the central truth underlying all biblical statements about God as creator . . . . [The world] is a product of God's wisdom . . . and exquisite care. It follows, then, that human wisdom necessarily has an "ecological" dimension, that is, it involves directing our own careful attention to learning about and preserving the world that God has made.[83]

The cantus firmus of God

We are within creation. We must listen deeply and attend to the wisdom that is the deep note, the *cantus firmus* of God. Keller says another name for this is Gaia, drawing on the work of James Lovelock—a word to express the connectivity, non-separability of all things, the entanglement that holds us all, and all there is.[84]

Christians don't tend to talk of Gaia. Such language suggests a female God, taking us towards debates about pantheism. But perhaps "Gaia" helps us think theologically about God's presence in the world, without compromising God's mystery. What we know of God and God's creative power is as "sound that goes out into all lands and words to the ends of the world" (Psalm 19:4), sound that never dies away. Sound and sense entangled, God with the world in love, as limbs not severed by fulfilment of desire. This brings us close to the attractive language usually used of Gaia.

Often, in current green thinking and writing, you will hear criticism of Christianity for leaving a legacy of dominion over nature, and behind this is the notion of a God over against the world, dialectically transcendent, wholly Other, who exercises power as domination rather than dominion, teaching humanity to do likewise. An article was written in 1967 by mediaeval historian Lynn White, claiming that the "dominion text" of Genesis 1:28, that humanity should subdue the earth and have dominion over every creature, was at the root of the ecological crisis because it licensed humanity to regard the earth as their possession, to do with it what they wanted, to exploit it, turning its resources to material advantage. Lynn White went on to say that it was important to draw on religion to counter this; but that's seldom remembered.[85]

There are many, many Christian poets and writers and activists today who offer more profound interpretations of God's creative power, and the ways humanity participates in that power, but their counter-arguments haven't dislodged the mindset of most (secular) ecologists. And of course, Christians—along with the rest of humanity—have been guilty of poor, no, sinful stewardship. But we have resources that can inspire us to live in hope, even as we struggle in the shadows of a bleak future.

I found myself, on *The Lark Ascending*, waking each morning alive to the world around, delighting, attending, with senses alive to the beauty of our environment. Aware, with thankfulness, of the beautiful world God has made and of my immersion in it.

Psalm 104 sang my soul

Psalm 104 is a song of praise to God the creator: God who can be seen in the light, and the wind: wrapped in light as in a garment, riding on the wings of the wind. And the myriad diversity of the created world is there: the leviathan, playing in the deep; the coneys, the animals of the forest, the birds of the air. And from the rich gifts of creation, humanity is satisfied. Meat, vegetables, wine, oil, bread. What more could humanity need?

This is a God whose love reaches beyond the farthest star; is deeper than the deepest ocean; is greener than the greenest green. A God whose glory fills the earth. A God, wrapped in light, as in a garment, who leads as in a dark cloud, a thick darkness.

For the darkness and the light are both alike to God

There are passages that remind us of God's wrath. The prophecy of Jeremiah, for instance:

> Take up weeping and wailing for the mountains,
> and a lamentation for the pastures of the wilderness,
> because they are laid waste so that no one passes through,
> and the lowing of cattle is not heard;
> both the birds of the air and the animals have fled and are gone.
> I will make Jerusalem a heap of ruins, a lair of jackals;
> and I will make the cities of Judah a desolation without inhabitant.
>
> *Jeremiah 9:10–14*

The threatening of the covenant between God and humanity causes the land to turn to desert, forest to be destroyed, rivers to run dry, crops and vineyards to fail, the animals to stop reproducing. When we read such passages (and the Psalms and other Wisdom literature) in the light of the reality of global warming, we hear a judgement upon us, a call to repent.[86]

I wanted to hear again the wild patience of God, who creates and sustains the universe, despite the human sin that desecrates the natural world; the sound of God's moral purpose—not only for humanity, but for the whole of creation, of which we are a part.

The palaeontologist Simon Conway Morris argues in his book *Life's Solution* that it isn't chance that the world is as it is. He talks of convergence in the natural world, by which he means that the emergence of our sentience as human beings was effectively inevitable, because the universe functions with an underlying pattern which makes certain outcomes predictable.[87]

Thou shalt not steal

That pattern, Simon Conway Morris maintains, has moral purpose. It is a moral pattern that lies at the heart of the relationship between God and humanity concerning the creation. It offers us a framework to take seriously our accountability to God for our stewardship of the rich resources at our disposal, within a universe that has moral purpose. He refers to the cosmic view of G. K. Chesterton:

> Reason and justice grip the remotest and the loneliest star. Look at those stars. Don't they look as if they were single diamonds and sapphires? Well, you can imagine any mad botany or geology you please. Think of forests of adamant with leaves of brilliants. Think the moon is a blue moon, a single elephantine sapphire. But don't fancy that all that frantic astronomy would make the smallest difference to the reason and justice of conduct. On plains of opal, under cliffs cut out of pearl, you would still find a notice-board, 'Thou shalt not steal.'[88]

We should not steal or exploit that which is not ours. Trespassers shall be prosecuted—if only that were true, as fires are set deliberately, each year, in the Amazon.

In the psalms that accountability is clear. It is expressed most particularly when things go wrong; when humanity is not wise in its relationship with creation. And so the psalmist expresses God's lament, anger even, at a sinful humanity that is careless and unwise in its treatment of the world:

> The earth trembled and quaked;
> the foundations of the mountains shook;
> they reeled because he was angry.
> Smoke rose from his nostrils
> and a consuming fire went out of his mouth;
> burning coals blazed forth from him.
> The Lord also thundered out of heaven;
> the Most High uttered his voice
> with hailstones and coals of fire.
> He sent out his arrows and scattered them;
> he hurled down lightnings and put them to flight.
> The springs of the ocean were seen,
> and the foundations of the world uncovered
> at your rebuke, O Lord,
> at the blast of the breath of your displeasure.
>
> *Psalm 18:8–9,14–16*

A picture emerges of God as faithful, true, and pure with those who are the same, but angry when people are crooked, with physical manifestations that can be read from the natural world. The earth trembles and quakes; the foundations of the mountains shake; thunder, hailstones, and coals of fire, lightning—all are the blast of the breath of God's displeasure. The anger is always secondary, though, to God's primary covenantal love.

It's important to read this poetically, metaphorically. There isn't a literal connection between God's wrath and natural disaster. It isn't true that each time there is a tsunami, or hurricane, God is visiting vengeance on the people who suffer. But yes, when the deep order and pattern of the natural world are thrown out of kilter, when the earth is no longer securely founded and becomes shaky, then there are consequences. God's engagement with creation is characterized by sustaining love, and the order and pattern of the natural world are a direct expression of that love. When humanity disrupts that divine order, then the psalmist is clear: God's wrath, disappointment, and anger are expressed, with dire results.

It's hard to continue as normal, when the enormity of the impact of climate change strikes home. We face monumental consequences, and as a priest I'm used to speaking of the love of God. It's much more difficult to talk of a God who is wrathful, angry, full of rage.

But not only that. For despite all the reasons for terror and despair, there is another lesson to be learned from the psalms—one that John Donne preached on, as reported by Ellen Davis. He took as his text:

> Because thou hast been my help,
> therefore in the shadow of thy wings will I rejoice.
> *Psalm 63:7 (KJV)*

Ending on a note of joy

Donne's sermon ended on a note of joy:

> But as in the face of Death, when he lays hold upon me, and in the face of the devil, when he attempts me, I shall see the face of God (for everything shall be a glass, to reflect God upon me), so in the agonies of death, in the anguish of that dissolution, in the sorrows of that valediction, in the irreversibleness of that transmigration, I shall have a joy which shall no more evaporate than my soul shall evaporate—a joy that shall pass up and put on a more glorious garment above and be joy super-invested in glory.[89]

Could I engage again with the joy that has always filled my heart and mind and soul, despite the despair and lament I felt? John Donne's fierce joy inspires hope, still. I remember a wonderful passage by Michael McCarthy in his book, *The Moth Snowstorm*:

> For one late April the blackcap was singing unseen, deep in a hedge, and it was joy-inspiring; and across the garden was the most gloriously flowering of the cherry trees, and that was joy-inspiring too. Then on a Sunday morning—I remember it precisely—the bird moved into the tree and began its song.

> I was struck dumb in amazement.
>
> Here was this God-given, blossoming snow-white tree, which was breathtaking in its beauty; and here was this God-given, breathtaking sound coming out of it. This tree, this tree of trees, was not just an astonishing apotheosis of floral beauty. It now appeared to be singing.
>
> The rational part of me couldn't cope. It was all too much, and it fell to bits. I had gone way past simple admiration into some unknown part of the spectrum of the senses, and there was only one possible response: I burst out laughing. And there, in the exquisite fullness of the springtime, was the joy of it.[90]

Somehow, the cloud that obscures God is a cloud of darkness—of fear and wrath entangled. It is also a cloud as a garment of light, of glory and joy.

Leaving the Fens

Bank Holiday Monday, a couple of days after the party: that Monday had been the hottest on record. We had retraced our route along the Great Ouse, through Denver Sluice, and Salters Lode, and along Well Creek, remembering to leave a donation—this time—for the lock keeper at Marmont Priory. As we drove along the waterway, dead fish were all around us. As Dieter Helm says:

> We go on tipping more and more fertilisers, pesticides and herbicides into our watercourses, and allow the silt to run off the land from intensive farming practices. Industry adds its pollution too. Add all the plastic and other rubbish that gets thrown in and you can see the results alongside any major river.[91]

He argues, in the twenty-five-year Environment Plan, that the polluter needs to pay the costs of this destruction. Such carelessness wouldn't happen so often, fouling up our watercourses.[92]

◆ ◆ ◆

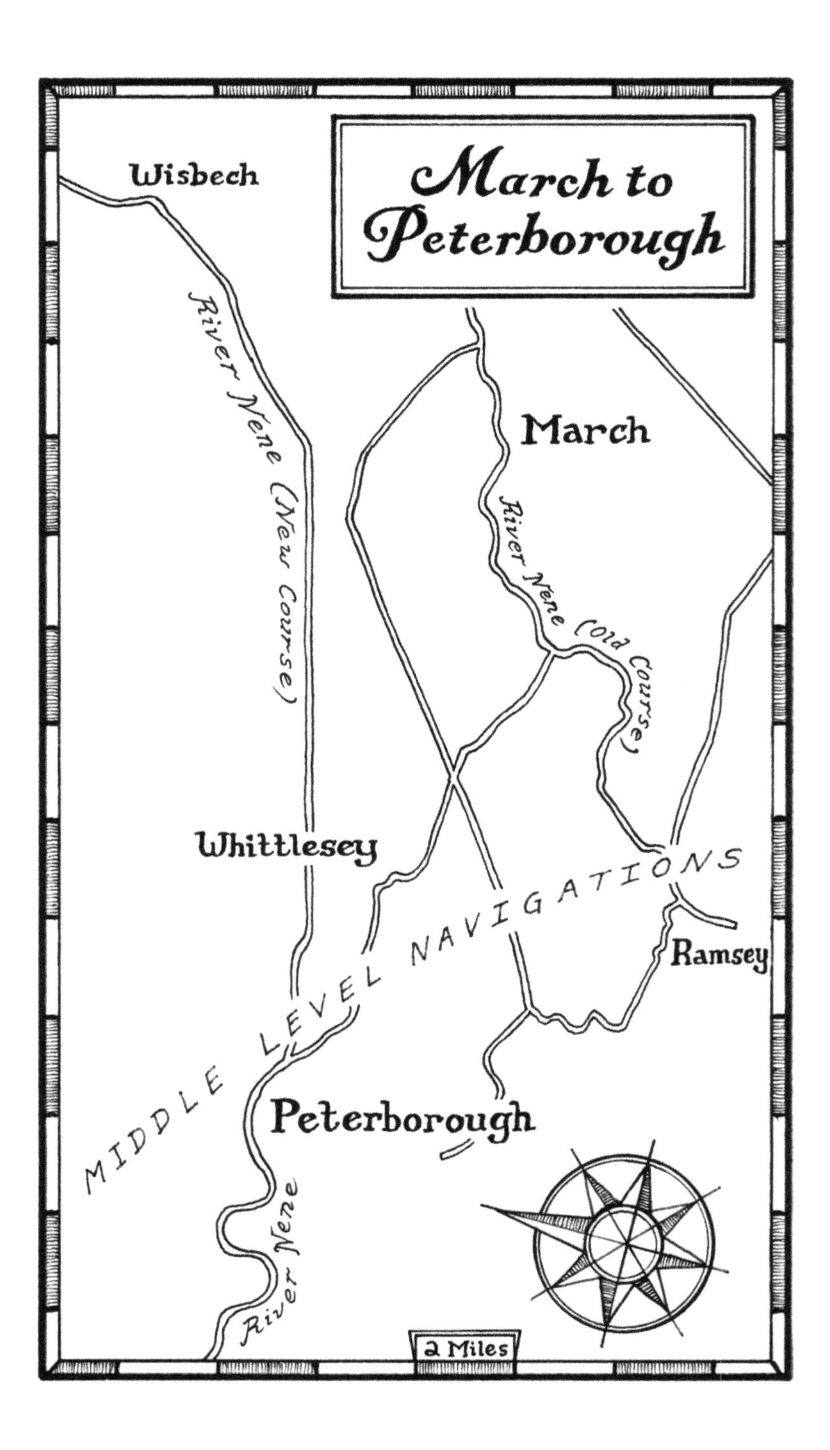

March to Peterborough
Wisbech
River Nene (New Course)
March
River Nene (Old Course)
Whittlesey
MIDDLE LEVEL NAVIGATIONS
Ramsey
Peterborough
River Nene
2 Miles

We were back at Fox's Marina. Once *The Lark Ascending* had been checked over, we had headed off up the Nene (Old Course), with steep banks on either side, through the narrow waterway that became Whittlesey Dyke, to Ashline Lock.

There we were told that there was no mooring space left in Whittlesey, and no mooring between here and Stanground Lock, on the outskirts of Peterborough. Undeterred, we negotiated the very tight bend (five-point turn) in the middle of Whittlesey and moored against the railing and concrete wall on Riverside Way, hoping none of the local residents would object.

The lichens on the wall were beautiful, exquisite circles of silver lace, so the air quality must be all right.

A walk into Whittlesey town, where there was one shop open, so we bought provisions and then settled down for a quiet afternoon and evening in the heat. The May blossom was just about to burst; birdsong was all around as I read Mark Cocker's *Our Place*. By that stage, I was deep in his analysis of the National Trust and its founders. Cocker argued that it had been hijacked by those concerned to preserve our heritage, rather than to campaign to save the wildlife and environments. His first chapter was all about his place, Blackwater, which he bought in 2012—five acres of floodplain in the parish of Postwick in North Norfolk—where he's restoring the dykes to fresh water from the dark, noxious sludge that comes with the encroaching woodland. Viv has a small property not far from Cocker's home, and told me she's doing the same with the sedge and ditches around her cottage. I said she should contact Cocker. They'd have much in common.

An entire way of life gone for ever

As Viv and I travel these dykes and waterways of the Middle Level, we are struck just how much work goes into sustaining the dykes and waterlands. It's a constant battle against the natural environment to keep the water at bay, in dyke and leam and ditch. I'm shocked, though, at the cost to the wildlife. As Cocker points out, the extraordinary natural abundance of wild vegetation and protein for the Fenlanders—reeds, sedge, herbage, flags, fish, ruff, plovers, godwits, cranes, herons, duck, geese, swans—was all lost as a result of drainage. An entire way of life gone for ever. It's true that our baseline is what we grow up with, and so it's hard to imagine the biodiversity and rich flora and fauna of past times—but as Cocker intends, it's important to remember, to inspire us to work for a different future when today's depletion is only a memory of a sad episode in this nature-loving nation's past.

He records a lament from 1620, as the drainage of the Fens began—"The Powte's Complaint" (the powte was a once-abundant fish). It's a song of protest, and was set to music by Patrick Hadley in his 1955 cantata *Fen and Flood* (Hadley's friend Vaughan Williams made a choral arrangement):

> Come, Brethren of the water, and let us all assemble,
> To treat upon this matter, which makes us quake and tremble;
> For we shall rue it, if't be true, that Fens be undertaken,
> And where we feed in Fen and Reed, they'll feed both Beef and acon.
>
> They'll sow both beans and oats, where never man yet thought it,
> Where men did row in boats, ere undertakers bought it:
> But, Ceres, thou, behold us now, let wild oats be their venture,
> Oh let the frogs and miry bogs destroy where they do enter.
>
> Behold the great design, which they do now determine,
> Will make our bodies pine, a prey to crows and vermine:
> For they do mean all Fens to drain, and waters overmaster,
> All will be dry, and we must die, 'cause Essex calves want pasture.
>
> Away with boats and rudder, farewell both boots and skatches,
> No need of one nor th'other, men now make better matches;
> Stilt-makers all and tanners, shall complain of this disaster,
> For they will make each muddy lake for Essex calves a pasture.
>
> The feather'd fowls have wings, to fly to other nations;
> But we have no such things, to help our transportations;
> We must give place (oh grievous case) to horned beasts and cattle,
> Except that we can all agree to drive them out by battle.[93]

The poem indicates the tremendous upheaval and dispossession of the enclosures of the commons that began in the seventeenth century and continued through to the nineteenth century. Fen drainage was how it happened on these middle levels through which we chug. I'm with Cocker in his passion to recreate the commons we once had, where something of the enormous variety of environments can develop again. He is worth reading, along with other nature writers today who all cry aloud for the same attention to be paid to the natural beauty of our national home, and who make economic arguments for sustainability.

Causing the Lord to lament

As we move along, with high dyke on each side, I continue to reflect on my faith, and what difference it makes—not worth having if it doesn't. I've always held that God's desire and will is for a better world, that a sense of common justice goes hand in hand with responding to the call of the love of God. The land around us continues to bear witness to the rapacious damage caused by our forebears, and by those who continue to extract as much as they can from the environment without footing the real cost at all.

You can hear the anger of the psalmist; the sense of injustice as strong enemies destroy the land, causing the Lord to lament. The springs of the ocean, the foundations of the world are all turned upside down, in *this* land around. I imagine the floods as Dorothy L. Sayers describes. How easy it would be to interpret God's hand of judgement, as the Fens reclaim their own:

> The springs of the ocean were seen,
> and the foundations of the world uncovered
> at your rebuke, O Lord,
> at the blast of the breath of your displeasure.
> He reached down from on high and took me;
> he drew me out of the mighty waters.
> He delivered me from my strong enemy,
> from foes that were too mighty for me.
>
> *Psalm 18:16–18*

I am left with my own sense of dry faith; how I'm not sure where I am—with God, with the Church. How this feels like a bottoming-out for me, a draining of all that I am, down to where the topsoil has blown away:

> I am poured out like water;
> all my bones are out of joint;
> my heart has become like wax
> melting in the depths of my body.
> My mouth is dried up like a potsherd;
> my tongue cleaves to my gums;
> you have laid me in the dust of death.
>
> *Psalm 22:14–15*

It feels God-forsaken

The usual comforts and veneer of life seem far away, as I steer this long boat through the drains of Whittlesey. It feels God-forsaken, the countryside around, and I find myself visiting God-forsaken places internally, where it's hard to justify the little I've achieved. Perhaps what Viv is doing is ultimately more meaningful, restoring a few acres of ancient fen, like Mark Cocker on his Blackwater. This was a once-watery land. Wicken Fen shows what's possible for more of it.

The fenland continued to haunt me, I found, as Viv, then Jenny and I continued our voyage to Skipton. The further up the country we went, the more our surroundings grew beautiful: less ravaged, less threatened. The scars that humanity had left were more obscure. As we voyaged the River Nene, then onto the canals, the negotiation of human with water seemed more benign, more respectful. Here in the fenland the negotiation felt brutal: the constant threat of the water a real and heavy presence, crying out for natural justice. I dreamt of bog and sedge, of butterfly and powte. I remembered times at Wicken; the sight of hobby, and owl, of hen harriers soaring, as once we watched.

I dreamt of the fens re-wilded over a greater and greater area, the reclaiming of the land from agricultural use where nothing but the crop can live to abundance, the return of dark and miry soil, harbouring amphibians galore. Where water, once again, fills parched land; where the dust of death comes to life, a resurrection of the earth.

I dreamt of the return of the hen harrier, now so close to extinction through poison, shooting, in the misguided desire to protect game. Here's a poem I wrote, "Harrier", on learning of the death of yet more breeding birds:

> The unforgiving growl of helicopter circling overhead
> round and round and round, seeking some prey, some pain
> to devour, leaves nothing in the air.
> It plummets from the sky –
> reviled raptor, the hen harrier.
> Graceful hunter of the sky kills again;
> viscerally loathed, destroyed, persecuted.
> The last confirmed sighting of the male harrier
> was on Friday 29th of May. Passing food
> to his mate on nest. Gone. And cold,
> abandoned, were the eggs in nest
> on first of June. That was Bowland.
> Geltsdale, Cumbria's breeding male gone too. The brood
> failed, 23rd of May. Gone from the air.
> We watched a hen harrier in the sky at Wicken Fen:
> its beauty sharp, nonchalant; with graceful play
> it performed, entirely in command;
> as if it knew we were entranced.
> To lose such grace I cannot understand.
> Ingrained habit to destroy will bring the day
> when they are lost and gone. Once gone, they will not come again.
> We cannot claim the skies are ours when all our waking dreams
> require supremacy, or else we do not know
> what lies behind, beneath, beside, beyond
> our loneliness and grief.
> Soul sick, sterile, we deplete the land
> to mirror our lack. Eden was far and long ago.
> Since then, instead of grace, ever-darkening regimes.

As we travelled on, through this fenland, our slow progress through these interminable, boring monocultures felt like a witness of sorts. We couldn't fail to attend—as you don't need to do, if you speed by in train or car—to the sadness of the soil, the dry sponge, this dry and thirsty land, crying for water. I saw the Lord as the psalmist did. A Lord who subverted the dominance of humanity hereabouts, the dominance of evil economic systems where neo-liberal market systems support the selfish interests of the fossil fuel industry, so determined to make a buck at the expense of the natural world and the future of the planet. The Lord who reigns over the water, come into its rightful inheritance:

> The voice of the Lord is upon the waters;
> the God of glory thunders;
> the Lord is upon the mighty waters.
> The Lord sits enthroned above the water flood;
> the Lord sits enthroned as king for evermore.
>
> *Psalm 29:3,9*

We were about to leave the Fenland Rivers and begin our voyage along the Nene, from Peterborough to Northampton. We would see the countryside around, instead of steep banks behind which the land stretches away, flat and drained, for miles on either side, with only the odd house roof and farm above the dyke. I yearned for such countryside, but vowed not to forget the Fens, and to look with eyes that have known this desolation. How has humanity made the earth? At what cost? What does God will and desire for this land? I ached with the psalmist:

> He makes me lie down in green pastures
> and leads me beside still waters.
>
> *Psalm 23:2*

I was ready to move on from the sterility that was myself, to find the goodness and loving mercy that would follow me, as a cloud, all the days of my life.

C H A P T E R 7

Peterborough and the River Nene

Stanground Lock marks the end of the Fenland Rivers, and the beginning of the Nene. The sight of Peterborough Cathedral in the distance greets your eyes as the guillotine lock gate rises, and a visit to the city and its wonderful church marked the transition.

The river is full and broad now, as we moor alongside the waterfront, and make our way up into the Cathedral Precincts.

It engages the eye with dramatic intent

It's an ancient site. There was a monastery here from 655, destroyed by the Vikings in 870 (the same onslaught that murdered St Edmund in Suffolk), and rebuilt in the tenth century. Hereward the Wake attacked the Abbey in 1069/70; there was a fire in 1116, and it was rebuilt in its present form between 1118 and 1238. The Gothic West Front is glorious—it engages the eye with dramatic intent, taking you from the depths of the three arches to the statues that gaze down upon you. The nave extends away, with a stunning diamond ceiling.

Rubem Alves is in awe of cathedrals:

> The cathedral: its vast, empty spaces, its silence, the light which fractures through iridescent glasses, the darkness which plays with the dance of the candles and gives movement to the stones . . . Yes, the cathedral is like the dead man: it is filled with unspoken words, filled with unexpected worlds . . . In it lives a Stranger—not the architect and his meanings—and when this Stranger speaks, one hears far away bells tolling inside one's soul:

> and the body trembles . . . Is the cathedral outside a metaphor of the cathedral inside? Or is it the other way round? I don't know—and the memory of an unknown home comes back to me. My knowledge fails me in that moment, I don't need it—I am possessed by the Wind which moves in the Void.[94]

Cathedrals do have tremendous power; they are full of void. Alves's dead man continues to haunt me—the dead man who is also the Christ, waiting for resurrection. The dead man of all past desires, bearing the grief and sadness, the guilt and pain.

The void is there, too, in Auguste Rodin's sculpture, "La Cathédrale".[95] He made it of stone, in 1908, and left the tool marks. It shows two right hands, and they make an inner space, an emptiness: the tenderness of anticipated touch; the *noli me tangere* of Christ.

I wrote a poem once, where I imagined Mary Magdalene whispering to Christ, sotto voce, as he lay in the tomb:

> Sotto voce
> I whisper to you
> the yearning
> of desire.
> Under my breath
> I breathe to you
> the murmur
> of my love.
> All that day long
> I dream to you
> the passion
> of the tomb.
> Touch me not
> O lover of my soul.
> Sever not
> the tense embrace
> that holds anticipation
> unfulfilled.
> Put not away

our life to be
presently
conceived of grace.

◆ ◆ ◆

I drop into the office to see if the new dean is around. He was Dean of Newcastle, and now is faced with the task of turning around the fortunes of Peterborough's cathedral. It hit the news a year or so ago, prompting a Cathedrals Working Group to review the leadership and management of these iconic buildings. Peterborough is a tough place to run. Like all cathedrals that rely on visitors, Peterborough Cathedral needs footfall; but there's not much reason to visit Peterborough—compared with other cathedral cities. He is on his way back from Leeds and will be going straight into a finance meeting, says his PA. The life of a dean today. I leave my best wishes. Later he emails me, sorry to have missed me.

We visit the grave of Katherine of Aragon, who passed away in nearby Kimbolton Castle. She died as she had lived, still convinced of the indissolubility of her marriage to Henry VIII, despite his divorce of her to marry Anne Boleyn, and his rejection of the authority of the Pope and declaration of his headship of the English Church. Elizabeth Cook captures the time brilliantly in her *Lux*, as she tells the story of Thomas Wyatt and the anxieties of that age.[96]

Katherine's burial there might have influenced Henry favourably to make the abbey a cathedral, and the abbot—John Chambers—the first bishop. Mary Queen of Scots was here too, briefly, after death, before her son James I removed her body to Westminster Abbey in 1612. We go on, tomorrow, towards Fotheringhay, where she was executed.

Back on board, we head upriver, following the narrowboat *Athena*, to the first lock at Orton. It's a stunning river, with green parkland and trees galore on either side. I hear—and then see!—a cuckoo. Another poem, "Cuckoo Egg", was inspired by a walk in the rain up Miterdale in the Lake District, when a cuckoo was calling, rather forlornly:

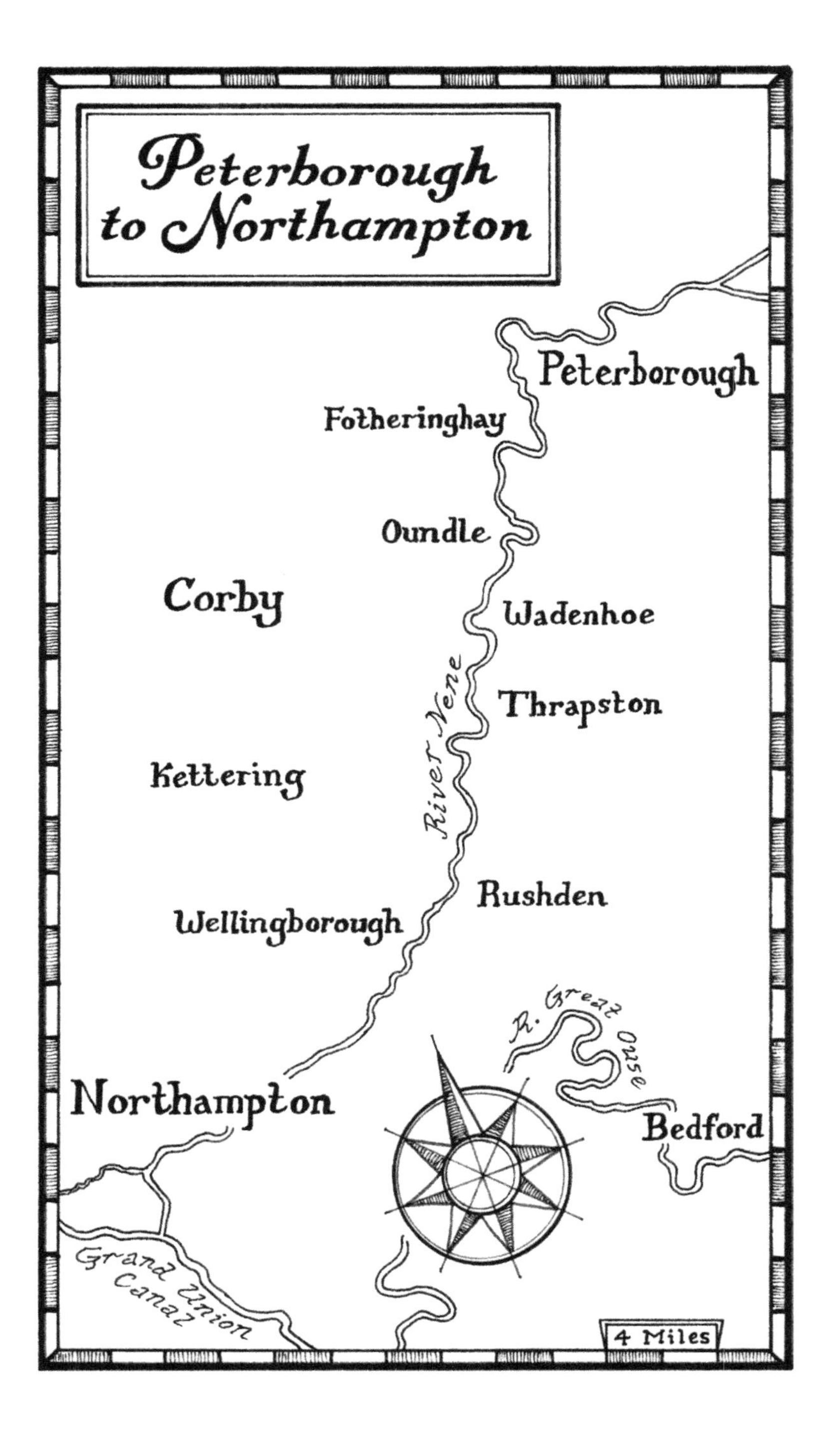
Peterborough to Northampton
Peterborough
Fotheringhay
Oundle
Corby
Wadenhoe
River Nene
Thrapston
Kettering
Rushden
Wellingborough
R. Great Ouse
Northampton
Bedford
Grand Union Canal
4 Miles

The jade is placed in the sideboard bowl –
a cuckoo egg. From larches, acid green,
the cuckoo calls its Maytime, careless echo.
It watches the chosen host lay her clutch –
dunnock, wagtail, robin, pipit –
the reed warbler, a favourite;
and then, with care, evicts one egg
to lay its polished, pale green,
blotched-olive egg; so cunningly
to match the warbler's four.
I heard a cuckoo as I walked in May
up Miterdale, in steady rain,
and on the path, a bone –
a skeleton nearby.

Common terns are fishing all around. Crested grebes, peewits, cormorants, moorhens—the birds seem more abundant than on the Fenland waters. I read in Mark Cocker's book that great crested grebe numbers fell to only thirty-two breeding pairs in the 1860s, owing to the popularity of their feathers for the hat trade. Five laws, passed between 1870 and 1880, offered the great crests the protection they desperately needed. In 1931 a census was carried out, known as the "Great Crested Grebe Enquiry"—one of the earliest and most successful counts, drawing in 1,300 volunteers. It concluded that there was a breeding population in England, Scotland, and Wales of between 1,234 and 1,241 pairs.[97] Numbers are now in the region of 6,000 breeding pairs. It's a gorgeous bird, with its strange crest, dramatic spear of a beak, and striking colours. We were too late to see a pair dancing their elegant courtship, but we did see adults carrying their young on their backs, tucked amongst the feathers.

Already England feels very different

Back on *The Lark Ascending*, we're happy to be alongside *Athena* as we tackle Orton Lock together. Its owners live on board, and cruise the river pretty much continuously. They show us the ropes, with helpful tips as we anticipate the thirty-seven locks of the River Nene ahead. They also warn us not to be too ambitious with the distances—Fotheringhay is six hours away. They advise us to moor up at Wansford. On we go to Water Newton Lock which is Downton-Abbey beautiful, with its mill and manor house, close by the church. Long lawns and the light stone of the houses, and already England feels very different, no longer the bleak desolation of the Fens.

The greens of early May are breathtaking along the Nene, as it meanders through meadows and woods. We push on to Wansford, where the mooring is already taken, and through the small town, mooring just below Wansford Lock, ready to go through first thing tomorrow. It's later than we thought, as we settle down to spare ribs, peppers and stuffed mushrooms—a great concoction by Viv.

The weather's turning—it won't be so hot tomorrow—or so we expected. We didn't know we were in for the longest, hottest spell since 1976.

As I settled to sleep, I reflected on the space cathedrals hold: the holiness that is so rare today. The air is alive with the breath of God, ever moving, ever still. Those verses from the psalms became a mantra in my mind and heart. I let them reverberate as I stilled my thoughts from their entanglements.

In God, in stillness

On God alone my soul in stillness waits;
from him comes my salvation.
He alone is my rock and my salvation,
my stronghold, so that I shall never be shaken.
Wait on God alone in stillness, O my soul;
for in him is my hope.

Psalm 62:1–2,5

> O God, you are my God; eagerly I seek you;
> my soul is athirst for you.
> So would I gaze upon you in your holy place,
> that I might behold your power and your glory.
>
> *Psalm 63:1,3*

I try to imagine gazing on God in God's holy place, but all I see is the rich velvet and red, gold, green colours of my closed eyelids. How do I know whether what I desire is what God wants of me? When do we know that what we discern is what's meant to be? Waiting on God, in God, in stillness. I'd had a deep sense of quiet hope as I'd sat in the choir stalls in Peterborough Cathedral. Surrounded by the power and glory of the place, all the history—and despite any present difficulties—I knew something of the power of the love of God. I knew I was here today, and gone tomorrow, but my life, my soul was infinitely precious to God, as are all created things, throughout the ages. As I drifted off to sleep, I felt God's love; the reassurance that "all shall be well, all manner of thing." Whatever might happen, nothing can separate us from the love of God in Christ Jesus.

The one loud noise that is made to the glory of God

Wadenhoe saw us moored up at the King's Head, alongside the boat *Echo*. Last night her owners told us that the next lock—Titchmarsh—was broken, so we might be a while on this stretch. We had our meal at the pub, then wandered up to the most delightful little church, dedicated to St Michael and All Angels, on a hill above the town. It looked beautiful from the river as we moved off on our way the next morning.

The bellringers were in that evening and were very welcoming, encouraging us to have a go. So we did. Dorothy L. Sayers writes, in the Foreword to *The Nine Tailors*:

> From time to time complaints are made about the ringing of church bells. It seems strange that a generation which tolerates the uproar of the internal combustion engine and the wailing of

> the jazz band should be so sensitive to the one loud noise that is made to the glory of God. England, alone in the world, has perfected the art of change-ringing and the true ringing of bells by rope and wheel, and will not lightly surrender her unique heritage.[98]

That was true of Wadenhoe—a village that had a strong sense of itself. The church appeared to be surrounded by the mounds of a mediaeval village—perhaps decimated by the Black Death? The living village is beautifully cared for, with a Wadenhoe Trust, and various small businesses in evidence—but no shop we could find. The pub got busier and busier as we returned to *The Lark Ascending*.

◆ ◆ ◆

I've found the best way to parallel park is by a gentle use of forward and reverse gear, with the tiller angled towards the bank. Every time we go through a lock, we need to leave the electronic guillotine in the up position, in case the river floods and the water needs to flow. As Viv does that, I bring the boat in on the upstream landing. Mostly I manage. At Lower Barnwell Lock I didn't. The bow swung out, and carried on swinging, caught by the headwind and the current of the river. With stern already secured, the boat ended up straddling the river across the top of the lock. Viv laughed and laughed. I wondered what on earth to do; and warned her if she took any pictures, she'd be dead. She did; she isn't. Thankfully there was no one around to enjoy my terrible plight.

The other six locks we did that day were uneventful, on a river where we saw only two other boats coming downstream. We're getting the hang of it—a routine which is beginning to look as though we know what we're doing. Except, of course, when the boat ends up wedged across the river.

The day had begun in drama too.

Viv's two dogs, Molly and Sid, are great company. Sid refused to come for a jog with me—little tyke—and as I returned him to the boat, Viv was shouting, as she lay in her nightie, stretched out trying to reach Molly, who had fallen in. Molly has only three legs, having lost one last September when a car hit her. Viv was worried that, as a result, she couldn't swim. She can, so we found—and with a bit of scruff-of-the-neck lifting ended up safely back on dry land.

I lost my Phoebe last summer. She was seventeen, and we'd had her since she was a puppy. I was reminded of the loss of her as I lived alongside Viv and her two dogs—how they responded to her kind and patient loving. So many boats have dogs—lurchers or terriers usually. Dogs that can fend for themselves, with their own resourcefulness. We often saw them running the length of the boat roof, entirely at home. Phoebe used to sit on the back of my kayak, as we paddled the river Esk in the Lakes, balancing delicately, wanting simply to be there, with me. Viv's dogs are faithful in the same way.

◆ ◆ ◆

We stopped for a break at Fotheringhay—disappointed that the church is cloaked in scaffolding. It is a stunning site, sailing majestically above the fields, with the river meandering below. It was built by the royal Dukes of York in the early fifteenth century. The lantern-tower, nave, and aisles were the work of William Horwood, a Stamford "free-mason". The choir, which was completed by 1415, lost its roof and was destroyed in the sixteenth century, so the church now has a strange, truncated appearance which accentuates the tallness of the tower. It was built at a time when practically all other church-building in the Midlands had ceased because of the agricultural depression following the Black Death.

We enjoyed a coffee at The Falcon, writing postcards to Peter—a continuous letter-to-be—and we climbed the castle mound where Mary Queen of Scots was executed, and Richard III was born. With views to die for, there are worse places to meet your end—or meet the world.

We were in Northamptonshire—but only just. Cambridgeshire and Bedfordshire are over the border and Bucks not far away. This is so different from the Middle Levels, or the Great Ouse. There are meadows and woods all around, and church spires everywhere you look. Prickwillow feels like an age away, in time and place.

Then the swifts arrived!

We saw five on Wednesday 9 May.

Swifts are enigmatic birds—they spend all their lives in the air, only landing to nest—hence their generic name, *apus apus*, "without feet". They even mate on the wing. There's that lovely scene in the 2007 film *Becoming Jane*, where the character Tom Lefroy reads from Gilbert White's *The Natural History of Selborne*, seducing Jane with his words. He describes swifts mating on the wing, falling until "the female utters a loud, piercing cry of ecstasy". Then comes a pause for dramatic effect, and the question, "Is this conduct commonplace in the natural history of Hampshire?"[99] It's a great moment.

Swifts feed on insects, particularly spiders, that float up on warm air. They bathe by flying slowly through rain. A swift covers a daily average distance of 500 miles, and the top speed recorded is 69.3 miles per hour, with only a diving peregrine faster. Each bird weighs only forty grams. The age of the oldest ringed swift is twenty-one years; the average UK life span is five and a half years.

They are not flourishing though—the rate of decline between 1995 and 2016 has been 53 per cent. The most significant factor in that decline is the destruction of nesting sites. Swifts are nest-site faithful, so it's devastating when the sites they return to are destroyed, or the holes blocked up, or . . . the church is covered in scaffolding. I really hope the scaffolding comes down soon on Fotheringhay Tower—because I bet that's a great place for the swifts to nest. There are ways in which swifts can be encouraged to nest in church towers: specially designed nest boxes that can be installed, as part of a national "Save our Swifts" movement, that wants to see legislation requiring all UK house builders to install swift bricks in all new-build homes, and incentives for retro-fitting nest boxes in older properties. The Diocese of Oxford has advice on its website.[100]

I love them—wheeling and screaming above, scything through the air. We'll put boxes up at St Michael's Workington, in hope for swifts. All churches should.

Another of my poems, "Ribcage":

Wild birds roost here.
Most days
a blackbird, whose sharp alarm
fires my pulse.
Too often
a crow, with harsh caw, takes me
to bleak ground.
I like it when
the dipper comes to stay, at home
in waterfall.
Best of all,
the swift flies through the bars.

Ascensiontide

It was Thursday 10 May—Ascension Day—and we were stuck at Titchmarsh Lock on the River Nene. "Where is Alan when you need him?" asked Viv—not usually one to turn to a man for help. Members of the Environment Agency were scouring the country in search of parts for the gearbox that operates the guillotine gate. Viv and I thought we might be stuck for days.

One of the few downsides of motoring along is the noise. I keep wondering about the alternatives. We simply couldn't afford one of the very new electric-powered boats that are now emerging on the market—but how wonderful it would be, to be silent as we move along! It's so hard to hear the songbirds. I've yet to hear a lark—even in the morning or evening when we've come to rest. I'd hoped that they would be singing all along the stretches of the Nene from Wadenhoe to Wellingborough that we travelled on Ascension Day.

We heard a distressing story as we went through the lock at Whiston. An old bloke, obviously a keen birdwatcher by his garb, remarked on our boat's name, and pointed out a field below us that was yellow and recently sown with some grain or other. He said the farmer had just sprayed the field with pesticide and killed off several nesting skylarks. Just like that. Gone.

Meadows: gone before we understand what we are losing

The more I read of Mark Cocker, the more I gather just how devastating the farming practices have been on wildlife over the last fifty years or so. He is arguing for a proper valuing and a strategic approach, like Dieter Helm who writes this in his *Green and Prosperous Land*:

> There are now virtually no natural meadows left in Britain. Some 97 per cent have gone, leaving only fragments in nature reserves. This is not just a tragedy for the plants that have gone, and the rich tapestry of colours that have been replaced by a uniform vivid green, but also for everything that depends on them. Evolution has produced very complex symbiotic relationships between the plants, insects, fungi, birds and mammals of the grasslands, and when the plants are gone, a host of these relationships collapse. They can be gone before we have had a chance to even understand what we are losing.[101]

Meadows—some ancient—have been destroyed since Edward Thomas wrote the poem "Haymaking" in 1915. Within thirty years two-fifths of them were gone. By 1960 another 1.75 million acres had gone. In 1984 just 3 per cent remained. Today it's 1 per cent.[102]

◆ ◆ ◆

We were stalled for a couple of hours at Titchmarsh. Not as long as we feared, thank heavens. Once through, we made good time from then on by pairing up with a couple whose work it is to transfer narrowboats from one part of the country to another. We went in tandem with *Luxury* on her way to her new owner at Milton Marina. They didn't hang about.

The benefit is mutual: both boats fit into the lock, so the work of raising and lowering the paddles is halved. Each boat takes it in turns to go ahead and prepare the next lock, waiting for the other to finish off reversing the lock (leaving the downstream guillotine gate up for the next), and catch up. So we went through ten locks—in order: Titchmarsh, Islip, Denford,

Woodford, Lower Ringstead, Upper Ringstead, Irthlingborough, Higham, Ditchford, and Lower Wellingborough. It sounds like a list poem.

◆ ◆ ◆

Ascension Day. I love the image of Jesus disappearing into the clouds. Perhaps this was the alpha and omega of the cloud of unknowing, the dark cloud that must be known if God's glory is to shine. If negative theology casts the shadow required for the sun to have its light, then we need this mysterious image. Just as Jesus was taken by the cloud and his disciples were terrified by the voice that came from the cloud, saying, "This is my Son, my Chosen; listen to him!", so we need clouds to overshadow us at times, in order more clearly to see.

The day of the Ascension. A day when it's appropriate to honour Vaughan Williams's work *The Lark Ascending*—which, as it happens, is Mark Cocker's choice to illustrate how Britons interact with wildlife. He explains how the post-war destruction of meadows, for land for urban sprawl, but also because of the intensification of farming practices, happened largely because of fear of vulnerability to starvation. Too many remembered the Second World War and were determined that it should never happen again. But since the 1950s, and particularly the 1970s, these have been decades of the upsetting of the ancient collaboration and partnership between land and people, explicit since before Shakespeare in poetry, literature, and music.

This is what Cocker says of *The Lark Ascending* by Ralph Vaughan Williams:

> It is worth recalling that its inspiration is a bird vocalisation from a species that is the most agricultural in Britain . . . [S]kylarks not only dislike woods: they are seldom to be found even close to a tree. The skylark is the quintessential inhabitant of ploughland or pasture, and our agricultural presence in Britain made its own abundance possible. Prior to the echo of a Neolithic axe upon our post-glacial wildwoods, there may have been no such thing as a singing skylark in this country.

> George Meredith's poem of the same name, which had first unlocked Vaughan Williams's responses, enlarged upon the shared ecology of bird and Britons.[103]

The imagination of that close relationship of human and nature is fired, religiously, on Ascension Day—the day when Christians stand and watch, their minds and hearts soaring as Christ returns to heaven in a cloud so that he might be eternally present on earth, and to prompt the Holy Spirit to come at Pentecost.

The thin line of sound that stretches to breaking, ever upwards, a dark point in the sky, which then parachutes down, to rise again: a wonderful metaphor for the Ascension.

There's a list poem I wrote a few years back, where I tried to capture in twelve short lines the life, death, resurrection, and ascension of Jesus. I'd been inspired by a collective noun: goldfinches together are a "charm". What other collective nouns could be used of birds, I wondered. And how might we see the narrative of Christ's life, from annunciation, to baptism, to teaching and healing, to Palm Sunday and Holy Week, Easter and Ascension as natural theology, told through the characteristics of birds? It's entitled "A Gospel of Birds":

> An innocence of doves
> A plunge of gannets
> A parable of blackbirds
> A balm of robins
> A fickle of sparrows
> A betrayal of magpies
> A condemnation of crows
> A torture of shrikes
> A grief of curlews
> A hope of swallows
> A sight of kingfishers
> An ascent of skylarks

◆ ◆ ◆

Viv leaves on Monday, after Jenny arrives on Sunday. Together the three of us will do the seventeen locks from Northampton to Gayton Junction, where we'll leave the River Nene, or "Nen" as it's pronounced around here, and enter the canal system.

We find where we are on a road map and compare with the canal map. Such different perspectives. It's weird to think that there's so much waterway activity going on that the road map or sat nav has no way of showing, as people speed past on the motorway. Your reality is how it's framed, so often. I'd no idea, when driving the motorways before I started to explore these canals, that this world existed. It doesn't just exist as place, though. It exists as time, perhaps more profoundly: slow and thoughtful time.

◆ ◆ ◆

My uncle died in Australia last year, very suddenly. He'd been a real character who, with Jenny, had developed a property outside Geelong. He loved boats. So when I was there for his funeral, and talking of my plans for this trip, Jenny agreed to join me for a couple of weeks. We agreed that Bill would have loved the thought; it seemed a good way to remember him. Jenny and I don't know each other very well, but we're not so far apart in age terms. My mother (RIP, 2005) was born in 1938, just before the war, and her siblings came after it, when my grandfather returned to Australia—all born within five years. She had me when she was twenty-one, so I was only twelve years younger than Bill.

Jenny knows a lot about water—it's been her profession: a lawyer in Australia, concerned with water rights and disputes. She's never been on a narrowboat before, so I'm glad Viv is with us for Monday to share the locks with us both. Viv and I had worked out a division of responsibility: I've minded what the boat is doing—bringing it into the lock, ensuring it's moored carefully and able to rise and fall with the water—while Viv has been the lock person.

I always thought Northampton was further north than this (but I guess in relation to Southampton it's where it should be). We've been travelling south-west since Peterborough, so Gayton Junction will see Jenny and me begin the climb northwards, going up the country, to some places

I've never been before. Canals, rather than rivers, too: a new experience for me.

So, back to Saturday, and after I booked into the marina, Viv and I took ourselves off into town. I was looking for a church to worship at on Sunday. St Giles' looks a little enthusiastic for me; All Saints' much more my cup of tea. For the Seventh Sunday of Easter, the Sunday after Ascension Day, the music of the mass is Leighton in D, and the anthem "*Notre Père*" by Duruflé. The rector is preaching and presiding. As I look at the notice boards, there's the Bishop of Richborough, smiling at the world, in a more prominent position than the diocesan bishop. OK—reading the signs—so this parish accepts the episcopal oversight of one of the "flying bishops". That means it's not in favour of the ordination of women to the priesthood or the episcopacy. It'll be interesting to see if there's any comment about the installation of the Rt Reverend and Rt Honourable Dame Sarah Mullally as the 133rd Bishop of London on the Saturday. It's always odd to encounter those who don't recognize my ordination to priesthood. It's as if some crucial part of me is invisible. I wonder if that's where I should go, and I decide I will. It'll be interesting to be under the radar.

Viv and I ate at *Pamukkale*, the number-one restaurant in town—delicious Turkish food and atmosphere. The name means "cotton castle" and refers to a natural site in Denizli Province in the Aegean region, in the River Menderes valley in south-west Turkey where the deposits of carbonate minerals have left white terraces. The ancient Graeco-Roman and Byzantine city of Hierapolis was built on top of the white "castle" which is 9,000 feet long, 2,000 feet wide, and 500 feet high. People have bathed in its pools for thousands of years. Pamukkale is now a World Heritage Site, giving its name to this restaurant. I think of the many UK citizens who holiday in Turkey (and other warm places) without a thought of the environmental costs of flying. How will that change? How hard it will be to give up those habits of holidaymaking we've learned, just since the 1970s, but which now are normal.

Cli-fi, it's now called

Then a performance of *The Importance of Being Earnest* was on in the Royal and Derngate Theatre, by the Original Theatre Company, so we went. Like so much traditional literature, its focus is on the relationships between people; Oscar's wit never fails, and was well delivered. Increasingly, though, it begins to seem irrelevant, unable to address the enormity of impending climate catastrophe. It's hard to lose yourself in the drama, with that constant sense of foreboding. But then, hard-hitting contemporary literature that faces into the threats of climate catastrophe—like John Lanchester's *The Wall*—isn't joyous escapism either.[104] It's an old genre, though—I remember Marge Piercy's book *Body of Glass*[105] and Dorothy Lessing's feminist sci-fi work, and the impact it had when I was a teenager. Cli-Fi, it's now called: it's a growing genre, climate fiction. A way to speak the despair.

Sunday morning, and I leave Viv on the boat with the dogs, to head off to All Saints'. A good choir, liturgy, and sermon. No mention of the new Bishop of London. No interest in who I was, or any particular warmth of welcome to me, a stranger in their midst. And again, the concerns that I'm dealing with—no mention. It's all too bleak; particularly when the Church generally is so concerned with its own decline and internal issues.

Viv and I had a rest day, cleaning and washing, pumping out, and filling up with water; reading the paper.

Lapwing

Mark Cocker had replied to my email letting him know that I was reading *Our Place* as we chug along. Hearing of Viv's property not far from him in Norfolk, where she's doing a similar job of clearing the marsh of trees, he encouraged her to get in touch. He said he'd heard of me from a mutual friend, Ronnie Blythe, who I last saw for lunch in September at his home on the Suffolk/Essex border. We agree that Ronnie, though getting forgetful, is still full of memories and stories to delight.

Mark explained some peewit behaviour I'd seen: a nesting male chasing off an interloper. His latest chapter that I'm reading is about peewits, or lapwings: how devastating has been the loss—65 per cent since the 1970s. The bird was of particular interest to him when he was a boy. He'd note the breeding pairs he could see from his bedroom window. He writes:

> What we know incontrovertibly now is that cold stains of absence have spread across the once solidly red map of lapwing distribution. Big holes have appeared across roughly two-fifths of central Norfolk. The worst losses, however, are in the western highlands of Scotland, much of Wales and in the English south-west, where almost the entire peninsula from Land's End almost to Bristol has been vacated.[106]

He says Lincolnshire, which uses the lapwing as the emblem for its Wildlife Trust, may well have to find another symbol. Like many other ground-nesting birds—snipe, grey partridge, turtle dove, skylark, yellow wagtail, corn bunting, yellowhammer—their numbers have plummeted. Increased use of pesticides, predation by foxes, intensification of farming (including silage making, just at the time of nesting, and when the young fledglings can't fly), are to blame. Unless there's serious remedial action, the losses will continue. Curlew numbers are hit particularly hard, as Mary Colwell explains in her *Curlew Moon*:

> The timings of agricultural activities have also been disastrous. Early in the spring, when birds are establishing their nest and beginning to lay their eggs, fields are flattened by heavy rollers and harrowed by chains to improve them for pasture and planting. Then, grass is cut for silage from late April onwards, when the eggs and young are still vulnerable.[107]

Cocker combines expert knowledge with beautiful writing. He speaks of loss, and how hard it is:

> For statistics and columns of figures do not begin to express the effects of the changes at a personal and interior level. For some people, agricultural intensification has triggered an emotionally charged, even visceral response, at the root of which is a baffling

> confrontation with local extinction and loss of meaning. The effect is powerful enough to alter an individual's personality and their entire view of life. It amounts to a persistent low-level heartache, a background melancholia, for which there is little remedy short of emigration.[108]

I know what he means.

◆ ◆ ◆

Jenny talks about Australia. That land has been so devastated by European human impact—as described by Germaine Greer in her 2014 book *White Beech*[109]—that it's almost as tragic as what's happening in the UK. Jenny and Bill tried to do something about it. Jenny told me about their property of seventy-three acres on the road out of Geelong towards the Western District, along the Barwon River, and how, with Bill, she made a real attempt to enhance birdlife and biodiversity. They planted trees which have attracted birds and wildlife.

Bill loved his water lilies. But this last summer and winter have been so dry that Jenny hasn't pumped water out of the river to keep them going—it just doesn't seem right. She described watching her sheep go off for meat, and whether she had the time and energy for sheep-rearing. Instead, she's going to let the land to a neighbouring farmer for merinos—that classic sheep of Spanish origin that produces such beautiful wool. Now marketed properly, merino wool is the best to buy.

Viv, Jenny and I sit out in the sun, alongside the boat, and talk away about how the last two weeks have been. Viv teases me unmercifully, warning Jenny of all my worst faults, which I deny vociferously, of course.

Jenny and I wander to look at Becket's Park Lock, so I can explain just how the paddles work, and where I'll position the boat, and what the routine is that we've developed. But we've got seventeen to do on Monday, and Viv is such an expert now that it'll be fine. Sally is joining us at Gayton, mid-afternoon, to take Viv and the dogs home. I'll miss them. But time ahead with Jenny will be really special.

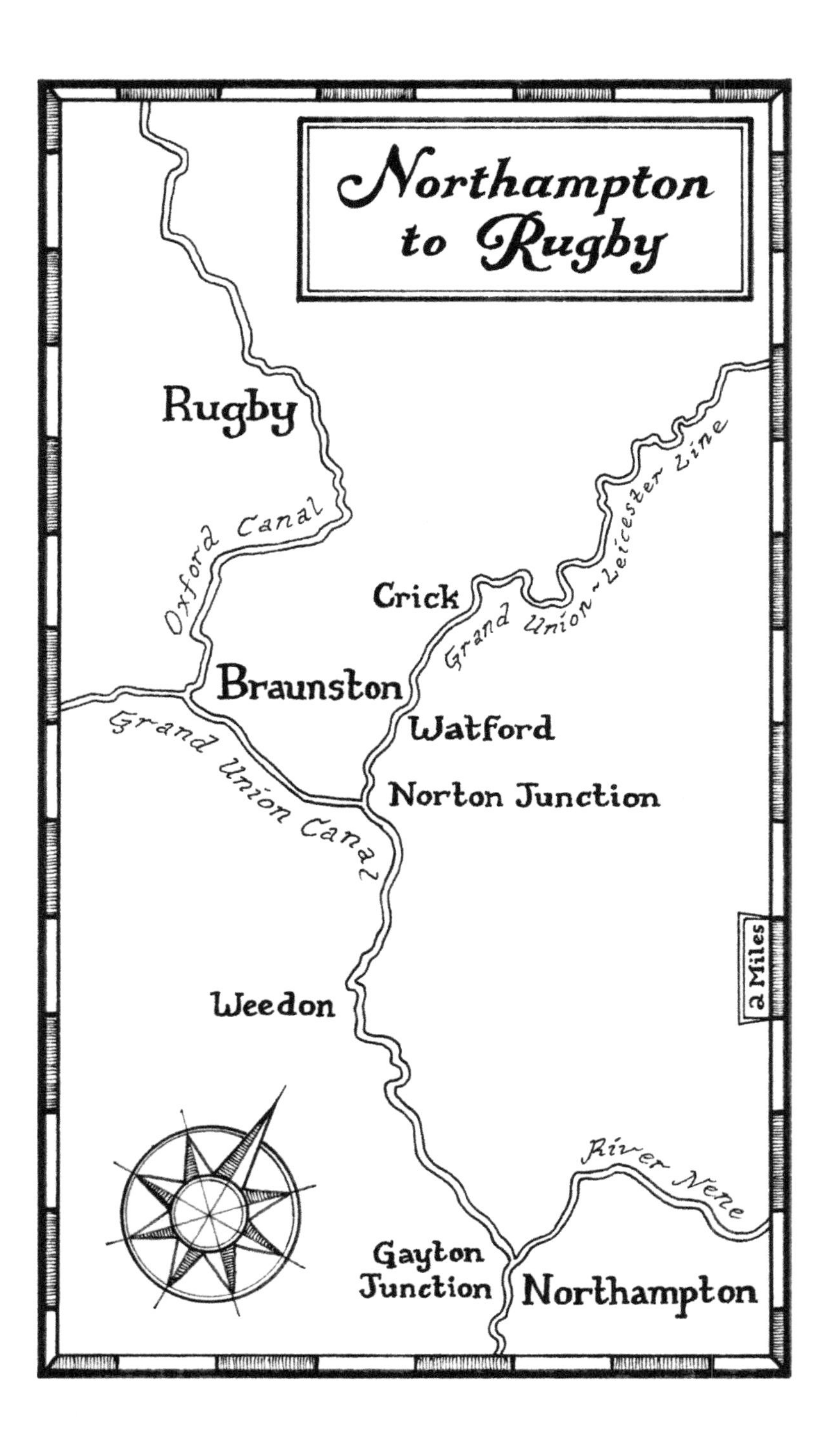
Northampton to Rugby
Rugby
Oxford Canal
Crick
Grand Union-Leicester Line
Braunston
Watford
Grand Union Canal
Norton Junction
2 Miles
Weedon
River Nene
Gayton Junction
Northampton

CHAPTER 8

Hawthorn Hedgerows Galore

Sally picked up Viv, Molly, and Sid from the marina at Gayton Junction on Monday afternoon. We'd spent the morning working the seventeen locks of the Northampton Arm of the Grand Union Canal, built to connect the rivers of East Anglia with the central canal system, and most significantly, with London and Birmingham. The Grand Union was completed by 1805, with the Northampton Arm added in 1815. The locks neatly fitted around the boat—room for just one—with vee doors at our stern, and one gate that swung open to let us out into the next holding pound.

We'd been warned not to head off too early, as these holding pounds can drain overnight—so we had a leisurely breakfast and made sure Jenny was settled in before we set off. All those locks later, and we're eating a late lunch of salmon, new potatoes, and asparagus at Gayton with Sally, amid much hilarity as we tell her just how impossible the last few weeks have been, what with my bossiness, and Viv's control freakery, and the dogs ruling everything. Even after they'd gone we couldn't get rid of them, as Viv forgot the dog food and had to come back. It was a lovely lunch, and a fitting send-off for the team that had been so brilliant over the last few weeks.

Then it was Jenny and me. We set off to take *The Lark Ascending* down the remainder of the Northampton Arm to join the Grand Union proper, turning right to head north towards Braunston. We moored up, after an hour or so, on a beautiful towpath outside Nether Heyford.

Viv hates Scrabble. So this evening gave Jenny the opportunity to thrash me.

Ferocious amounts of water

Tuesday was a long day, with thirteen locks, and the Braunston Tunnel to negotiate. The locks were serious—by which I mean deep and large—and the first couple saw Jenny and me turning the paddles up and down with ferocious amounts of water descending on us from high above. The trick is to find another boat to pair with, and thankfully the pair behind us decided it would be quicker if one of them came with us. So Mark and Caroline joined us through the remaining five locks and then gave us good advice for the tunnel ahead.

You need waterproofs for Braunston Tunnel, and your head lamp on, along with all the lights in the boat. It's dark, and dripping, and hard to see anything—except the bright head lamp of a boat approaching from the other direction. There's only just room for boats to pass, with much scraping against the sides of the tunnel. I have a small brick souvenir.

The tunnel, built by Jessop and Barnes, is 2,042 yards (1,867 metres) in length, and opened in 1796. It really is incredible to think of the construction. I'm full of admiration for the engineers who conceived that such a thing was possible. How would you begin to plan such an enterprise? Its progress was delayed by soil movement, and probably that caused the tunnel to have a slight "S" bend. There is just room for two seven-foot-beam boats to pass. We're six feet, ten inches wide. There are three air shafts along its length. I think of the enterprise and inspiration that spoke of confidence in the future. Yes, driven by the desire to accumulate wealth, I know. Somehow, though, we need to stir that same ingenuity now, to relinquish and restore in order to survive. Passing through that deep tunnel—our first—which took about twenty minutes, the focus is always on that small light ahead which never seems to get any bigger. You wouldn't want to fall overboard, nor have engine failure. How did they manage in the day of the horse? The silence. Dripping water. The occasional shout of warning or greeting. Then just eerie, closed-in blackness that spoke to my internal condition. I was glad to have to concentrate so hard on steering, and found myself reflecting on how dark depression can be transformed by attention to other purpose, particularly with a physical engagement. How different this is from the trivial, shallow gratifications of that neo-liberal system that wants us distracted from the deeper malaise. After this voyage is over, how can I transform the darkness inside into action and change for a hopeful future? What might that look like, physically?

Once we're through, though, there are other concerns, other locks to negotiate. There's always the next one to anticipate. We left Mark and Caroline as they moored up and we headed off for the next set of six locks, this time pairing with a boat with two couples on board, so lots of help for Jenny with the descending locks which took us into Braunston.

Braunston Junction is where the Oxford Canal meets the Grand Union at a three-way junction. This is the mecca of the canal system. There's lots to see, not least the beautiful Horseley Ironworks Bridges. We became increasingly aware of the industrial history—of iron and its production—as we moved further north. Braunston, the village, is on the hill above, where Jenny I had a cider at the Old Plough, but the life really happens along the canal with chandlers, and marinas, and boats galore.

What these canals once were

We passed a plaque to the Bray family and, intrigued, I googled them. Arthur Bray's obituary in the *Guardian* came up. It gives a brilliant picture of a life now gone—but also of what these canals once were, before the leisure industry revived them.[110]

Bray died in 1998 at the age of ninety-three, having been one of the last of the old working canal boatmen, the head of one of three families from Braunston "who together worked the last fleet of six paired narrow boats, carrying cargoes under regular contract." His life, spanning the twentieth century, told the story of the decline of the world of working the boats.

He'd been born in a boatman's cabin on the Grand Union Canal, at Norwood, in West London, on Christmas Eve 1905. Despite a complete lack of schooling, he knew how to judge the weight of a load, and how to calculate its worth in money. It was his job, from the age of five, to lead the horse. The Grand Union is 137 miles long, from London to Birmingham, and he knew every foot. By the time he was eleven, he was working a boat with another eleven-year-old—just the two of them on the Coventry Canal, unloading twenty-five-ton cargoes of stone for a quarry.

He worked through the Depression, through the transition from horse to engine in the 1930s, then through the Second World War, exempt from call-up, now married to Rose. The brief canal renaissance of the war years didn't last, for as soon as the motorways were built (particularly the M1 into the Midland coalfields), the pressure was on the boats. The winter of 1962 froze everything—no coal could move through the canal system, at a standstill for twelve weeks. The Brays worked, by this time, for Samuel Barlow, who had diversified into lorries. He sold the remaining boats to the new Blue Line Cruisers. As the *Guardian* article has it:

> These latter days were among the most tragic of the canals. To remain competitive, rates were low, and the boatmen were only paid for delivered loads. The three remaining families worked seven days a week, 15 hours a day, all year round to save their way of life. In the mid-1960s, for a week's round trip of 243 miles and 186 locks, a family shared a wage of £15. If they did not work, they did not get paid.[111]

Arthur Bray eventually retired in Braunston, still living on a boat until he went into a nursing home in 1996. His was the last working canal run. Hints of the past he lived are all around in the trappings of the leisure industry that now drives the boats. Easy to romanticize it all, to forget just how lives and landscapes are shaped and destroyed by economic forces. There was more evidence as Jenny and I turned up the Oxford Canal and moored on the towpath amidst cow parsley, surrounded by fields revealing the ridges and furrows of the peasant farming that endured from the Middle Ages until the enclosures.

Ridge and furrow

Each ridge and furrow would be an allotment for a family. When the hawthorn hedges were put in by farmers to enclose the common land into fields, whole settlements were displaced, often forced towards the burgeoning industries of the cities and their factories.

Many landowners became rich through the enclosure of the commons, while many ordinary folk had a centuries-old right taken away. W. G. Hoskins explains:

> Where such landlords, lay or monastic, owned the whole or greater part of the manorial soil, the eviction of the open-field farmers was easy enough. At the end of the farming year, immediately after the corn-harvest, they were ordered to go; their farmsteads were demolished; and the multitudinous strips of the open fields were laid down to grass. The two or three arable fields were replaced by a number of large pastures, enclosed by a hawthorn hedge and a ditch.[112]

An anonymous protest poem from the seventeenth century summed up the anti-enclosure feeling:

> The law locks up the man or woman
> Who steals the goose from off the common,
> But lets the greater felon loose
> Who steals the common from the goose.

So great was the need for hedges during the enclosures that a whole new industry sprang up supplying hawthorn plants to be used in creating new ones. Those same hawthorn hedges are now absolutely stunning—covered in heavy white blossom—clouds of beauty. In spite of all their current glory, they spoke of the past loss of a way of life; of people on the move, refugees driven by the greed of others. Things hardly change. Those patterns of life were so much more aligned with the created order; but how possible is it for whole populations in the world today to live that close to the land? Jen and I talked about how we must attend to our own food production much more responsibly. Eating seasonally with hardly any meat or fish is a start.

Lud-in-the-Mist

I remembered how Hope Mirrlees captured the glory of the hawthorn in her strange 1926 fantasy novel *Lud-in-the-Mist*. She describes the world waking up in the spring, then going over into dull summer. She marvels at the hawthorn, giving the words to the enigmatic Endymion Leer:

> Think of an autumn wood, or a hawthorn in May. A *hawthorn in May—there's* a miracle for you! Who would ever have dreamed that that gnarled stumpy old tree had the power to do *that*? Well, all these things are familiar sights, but what should we think if never having seen them we read a description of them, or saw them for the first time? A golden river! Flaming trees! Trees that suddenly break into flower![113]

Lud-in-the-Mist is a novel that McGilchrist would relish, I believe. It's about a town that has lost its dreaming, its right-hemisphere engagement with the Other—which is represented by the fairy land that lies to the West. To eat the fairy fruit that is smuggled in is to see the world in very different ways, ways that are extremely threatening to the powers-that-be, who want only that the world should be kept in control, on solid ground, known and without mystery. It is an allegory for any culture or society that becomes dominated by a left-hemisphere attention to detail and process, sacrificing the engagement with the Other that is so necessary to what it means to be human. And now we are caught by consumerism, rather than having an economic system built on sustainable, seasonal provision. Our profligate summer lifestyles can only have an autumn and winter, as we look to the bleak futures that lie ahead.

To see the may blossom lining the canals spoke to me of the impulse to control the land, to parcel it up for commercial gain, whatever the consequences for those who had made a living from the commons since time immemorial. But now, explosive in white, pink, red—the blossom is exuberant, a miracle. It tells of God's regenerative grace. I struggle with the delight, now always thorned with fear. We are so entangled in systems that have taken us far away from the land into global supply chains that are rapacious and leave us vulnerable.

I try to hold the delight and defend myself against despair, focusing on other flowers, modest on the towpath: forget-me-not, bugle, speedwell, comfrey, cow parsley. The canals, towpaths, and banks are wild highways, ribbons of silver edged with lace.

There is exuberance as the psalmist sings:

> Let the fields be joyful and all that is in them;
> let all the trees of the wood shout for joy before the Lord.
> For he comes, he comes to judge the earth;
> with righteousness he will judge the world
> and the peoples with his truth.
>
> *Psalm 96:12–13*

The enduring seasons have real power of regeneration. I try to still the anxiety and rest instead in the indescribable abundance and fullness of nature around, delighting in trees, flowers, even the grass that shouts for joy before the Lord. How, I pray, can I stay with this joy instead of always, it seems, falling back to the default of anxiety, to fearful lament for the future? I struggle to contain the despair within. It would help to do further research, when the voyage is over, on current economic and political thinking that challenges the all-pervasive neo-liberalism that has so shaped the habits of our lifetime. To turn this lament to flame, a fierce hope, away from that all-pervasive consumer mindset, to relinquish and restore a quality to life that might sustain us onwards.

The very middle of England

Onwards, now, towards Rugby and Coventry along the Oxford Canal, and then onto the Coventry Canal. There we were, in the very middle of England.

After the busy canal junction that is Braunston we encountered the three locks at Hillmorton (outside Rugby, on the map). Like Malcolm Guite, I noticed the poetry inscribed on the bar of the lock gates. He wrote it thus in his weekly column "Poet's Corner", in the *Church Times*:

> Our passage through the locks meant that we approached the phrases of this public poem slowly, gently, each phrase on its own at first, and then finally, and in retrospect, began to put them together: not a bad way to approach all poetry.[114]

The poetry was part of the project "Locklines", celebrating the inaugural year (2012) of the Canal and River Trust. Written by Roy Fisher, the four lines at Hillmorton (in the right order) are:

> Working water
> Captive for a while
> Climbs carefully down
> The door makes depth

Guite continues to explore the paradox of the water through locks. "Water," he says, "must always flow downhill, and yet here it was lifting us up. Detained, held captive for a while on its slow descent, it enabled our ascent, and the lock itself spelled freedom, the power to move in a new direction."

These locks were different from any other we encountered: two singles alongside each other to speed up transits. We've noticed differences in the locks as we've changed canals—these ones were deliberately designed at seven-foot-wide to restrict broader-beamed boats from other canal systems wanting to take advantage of the trading arrangements of this region. Boats were built at a certain size for certain waterways to control the business. The stop lock at Hawkesbury Junction, with its small step of water, prevents too much flowing from the Oxford into the Coventry. Even water was property.

It's only with the leisure industry that boats have been designed to travel the whole system—and *The Lark Ascending* is the optimal length and breadth at fifty-seven feet long, six feet, ten inches wide. (Though I've been told that the locks on the Leeds Liverpool will take a fifty-eight feet only on a diagonal, so fifty-seven will only just fit.)

It was windy this day, which always makes the boat difficult to handle. You'd be surprised at how hard it can be to moor. Mark and Caroline (from the day before, who we'd accompanied through the tunnel) joined us, which helped. It was cold too. When we reached Newbold-on-Avon, having passed Rugby with nowhere to moor, we decided to walk back into the town, not realizing how far it was by foot. We were looking for a bookshop and were disappointed to find that Rugby has no Waterstones. The town centre was rather tired too, but the church was beautiful.

A great welcome pack for visitors, as we entered. St Andrew's obviously has a good town-centre ministry, with a café in the church open through the day, and lots of groups active. The church has two peals of bells, one of eight, cast in 1895 at Whitechapel, in the North East Tower, and another of five, cast in 1711, in the West Tower, so St Andrew's is the only church in the world with two peals that are regularly rung in the English full-circle tradition. There was a good little explanation about change ringing too.

Walking into Rugby and back from Newbold took some time out of the day, and I was anxious about the loss of time, wanting to get as far as possible while Jen was still with me. I need not have worried, though. We would have been delayed anyway. On we went, to find, after the next bend or two in the canal, boats lining the towpath. Someone from the towpath calls across. "You won't go much further. There's a large oak tree across the water."

Brinklow, oak down

So we too moored up—near Brinklow—and went to investigate. It was a beautiful spot, in the middle of a wood. The news was that the team were on their way but were stuck on the M1. When they arrived, they set to with chainsaw, finishing at nine o'clock that night, with half the job done. This Eastern European team did a great job, and in the morning another team was there.

The falling and felling of the tree attracted a lot of attention. Over fifty boats eventually queued up each way, all with opinions. There was Nick Sharpe, Carrier, with his working boat, who got busy with his wheelbarrow, taking the sawn-up logs to his cavernous vessel—a boat that harked back to the hauling days of the past, when the boatman would have his cabin at the stern and the whole boat would be an empty hull to fill with goods. Then his barrow got a puncture, and he couldn't find the green slime so had to mend it with a bicycle repair kit. He told me he owned the boat, making just enough to cover the reduced working boat licence of £1,000 a year. He was off to Fradley for a job he was meant to do on Friday, but he didn't think he'd make it. He wasn't too bothered. These things happen.

There were opinions about why the oak had fallen, with a general consensus that it was rotten at base. A beautiful tree, though—a shame to see it down. I counted at least seventy rings. The oaks are at their best just now, glorious, alongside the may blossom—that bright, clean green which will soon fade to summer and into the tiredness of August.

There's some sign of ash dieback—the telltale orange on the trunk, and leaves missing from the outer branches and twigs—but on the whole the canals are great corridors of wildlife, in what is sterile arable land around.

Trees to capture carbon

There's research out now arguing that planting trees is absolutely the best and cheapest way to capture carbon from the atmosphere. Trees absorb and store carbon far more effectively than any technological technique. Professor Tom Crowther of the Swiss Federal Institute of Technology in Zürich (ETH) argues that trees could be planted on about 11 per cent of all land, equivalent to the size of the US and China combined. "What blows my mind is the scale. I thought restoration would be in the top ten, but it is overwhelmingly more powerful than all of the other climate change solutions proposed," he says. It's crucial to reverse emissions from fossil-fuel burning, and reduce current forest destruction to zero, and then forest restoration would take fifty to a hundred years to have its full effect of removing two hundred billion tons of carbon. But it's easy to do—it's available, and individuals can just do it. Jean-François Bastin, then also at ETH Zürich, called for political action: "Governments must now factor [tree restoration] into their national strategies."[115]

Trees are restorative in so many ways. It's hard not to feel joy when walking or sitting deep in a wood or delighting in the myriad hues of green. Green: the colour of hope. The stretches of canal that were tree-lined, or took us through woods, had their own silence, lovely, dark, and deep. Regenerative, with an ancient, holy peace that resonates through the rings of the hardest wood, the green-black of shade, the bright gold-green sunlit, insect-filled glade.

Yet again and again I was pulled back to compare the degradation of the countryside, poor as it is compared with my childhood in the 1970s. No voles, or stoats, or corn buntings, yellowhammers, flycatchers. Very few swallows; no more swifts since those sightings at Northampton. A few fields full of buttercups and other flowers, but largely the flowers are alongside us—meadow vetch, forget-me-not, comfrey, bugle, speedwell. Meadows should be full of them. Insects too.

Insect depletion

Research that tells us insects have suffered a 70-per-cent depletion is very serious. We can't do without insects. The lack of meadowland doesn't help. Those enclosed pastures, the ones left for hay, were of real ecological importance as open, sunny areas that attracted and supported abundant wildlife, providing space for courtship displays, nesting, food gathering, and a wide array of wildflowers, of utmost importance to pollinating insects, including bees, and hence the entire ecosystem.

Jenny and I walked into Brinklow and bought some pansies, which I later planted out in a tub decorated in traditional style that I bought from another boat. When we got back, the first boats were moving. The tree was cleared, and we were through by 11.15, with tight congestion continuing for some time. Goodbye, Brinklow.

The Oxford Canal took us on, then, to the outskirts of Coventry where we joined the Coventry Canal, through Nuneaton, and on to Atherstone, where we moored up. There were eleven locks to do the next morning, Friday.

A drink at a pub—not the best pub, as we discovered as we walked back, but never mind. We stopped to talk to a professional fisherman about how many ducklings are taken by pike. He reckoned a fish could take one a day. The largest brood we've seen was fifteen ducklings.

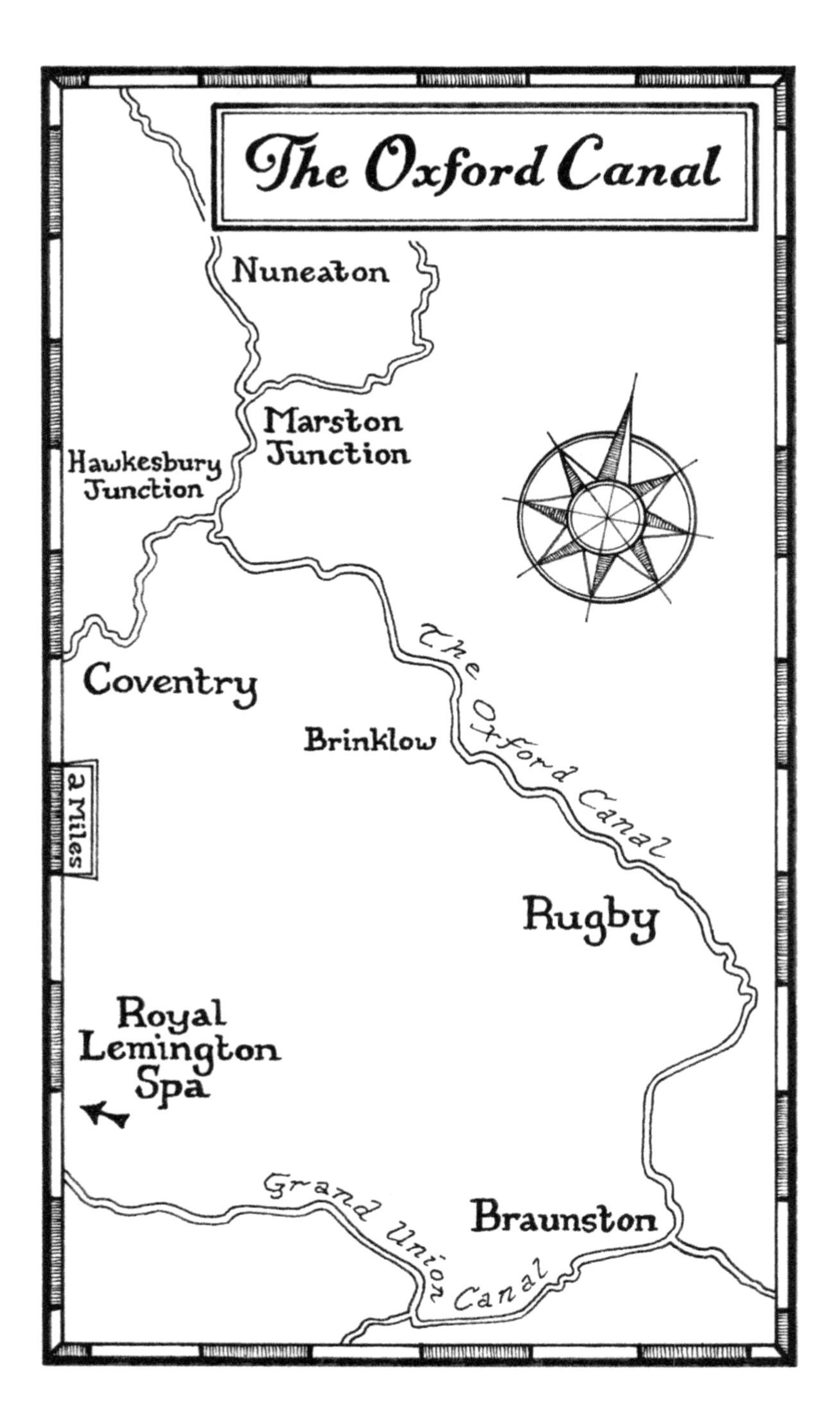
The Oxford Canal
Nuneaton
Marston Junction
Hawkesbury Junction
Coventry
Brinklow
The Oxford Canal
2 Miles
Rugby
Royal Lemington Spa
Braunston
Grand Union Canal

I've finished Mark Cocker's book *Our Place*. He argues that the lack of coordination amongst environmental groups must be addressed, if concerted action is to be taken to reverse the intense reduction of wildlife in the UK today. There are so many vested interests stacked against the flourishing of biodiversity, not least those of farmers and landowners, and the subsidies they receive to farm intensively. As Dieter Helm argues in his *Green and Prosperous Land*, a much better political engagement is exigent to save and nurture our natural capital.

The myth that everything's okay

One of the problems is the myth that everything's OK—that we're still a green and pleasant land. Though Cocker is good on what he describes as a "deep melancholia" that many feel—and how the natural world around us has for centuries replenished our collective spirit, by "immersion in nature's unfathomable and obliterating otherness, so that it can purge the travails and toxins of our own making". He writes:

> Nature's great and irreversible continuities—the passage of the clouds, the turning of the seasons—measure all our smallnesses. They put things in perspective. They render us humble. . . . Hope is written into all our connections with the rest of nature, and it is a two-way process. . . . It begins the moment you open the door to go outside. You have only to have the sun on your back, the wind in your face and birdsong in your heart to know their rivet-bursting powers of liberation. . . .
>
> It explains why we cling so tenaciously to the myth that this country continues inviolate. We don't want to hear that our final redoubt, the place where we go when our human condition is overwhelming, is itself in need. Alas, it is. In the twentieth century, the British drained their landscape of wildlife, otherness, meaning, cultural riches and hope.[116]

He continues that we are in denial:

> [W]hat we have done to our country becomes the truth that dare not speak its name. However, hope lies, surely, not in perpetrating any myth, not in doctoring the facts, but in owning them squarely and with the whole of ourselves.[117]

He insists that to save Britain's wildlife before it is too late requires a political and strategic approach. His book stirred me to wonder about how much more Christians can contribute. For natural beauty is one of the ultimate gifts of a creator God, and with our diminishing belief in that God goes our increasing disrespect and contempt for the natural world we are given. The rot is at the roots—just like that sad oak tree at Brinklow.

◆ ◆ ◆

Psalm 104 is a hymn of creation praise. It captures how God can be read from the natural world. We are held as part of all that surrounds us, connected, entangled with every atom of the physical world around. Water, with its constant flow, is at the heart of the fecundity of tree, animal, flower—all there is. Here it is the streams and rivers:

> They rose up to the hills and flowed down to the valleys beneath,
> to the place which you had appointed for them.
> You have set them their bounds that they should not pass,
> nor turn again to cover the earth.
> You send the springs into the brooks,
> which run among the hills.
> They give drink to every beast of the field,
> and the wild asses quench their thirst.
> Beside them the birds of the air make their nests
> and sing among the branches.
> You water the hills from your dwelling on high;
> the earth is filled with the fruits of your works.
>
> *Psalm 104:9–14*

Profound joy, mingled with sorrow

As the water pilgrimage unfolded, with the intense re-engagement with the natural world that canals offer, I wondered how the Christian religion might be brought back into the conversation. How might those deep passions and this peace be stirred, the yearning for God, at the heart of creation? The profound joy, mingled with sorrow, that comes with the delight of birdsong, the vibrancy and colour around, inspires me to write it out, this sense of the otherness of the world around that shapes us as we know ourselves in harmony and peace with the deepest impulses of the natural world, and as, through the beauty, we are made aware of the ultimate otherness that is God.

How fundamental is the poetic spirit to this knowledge of God! Here is another of my poems, "Bullfinch":

> When I was nine I bridged the barbed wire fence
> with our blue and red metal slide
> a trespasser into the orchard.
> There I opened, one after the other
> the farmer's cages freeing
> trapped bullfinches
> amidst the apple trees and blackcurrants.
> The hot scent of currant blossom, the delicate
> white pink of apple bud and that deep raspberry
> pugnacious breast thrilled me—
> I flew free, ravished by the plenty.
> Today I sit on beech branch swing, still
> and quiet as it rains large splats,
> watching a pair, now rare (amber status).
> They whistle and pipe their mournful cry
> at home among the saplings.
> They have the space, bullish with
> oblivious grace.

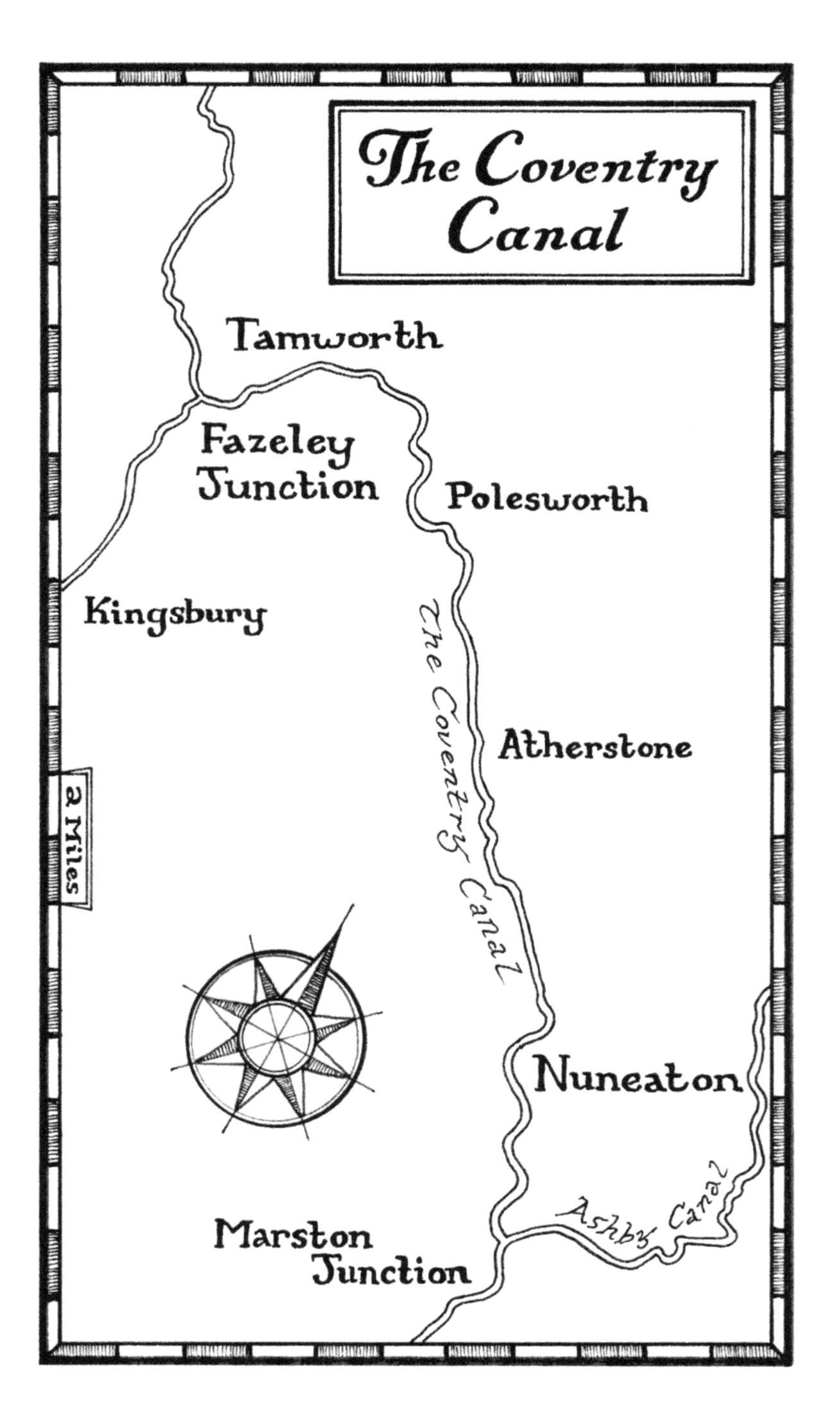
The Coventry Canal
Tamworth
Fazeley Junction
Polesworth
Kingsbury
The Coventry Canal
Atherstone
2 Miles
Nuneaton
Ashby Canal
Marston Junction

CHAPTER 9

Brindley Country

No TV for us on Royal Wedding Saturday. Jenny's really not sure what all the fuss is about. Perhaps we'll stream it later.

Friday saw us through thirteen locks, travelling for nine hours, broken only by a stop for breakfast at Polesworth, where we had eggs and bacon at a little café before exploring the Abbey Church of St Editha. Polesworth has, also, an antiquarian bookshop. I came away with two volumes of Richard Hooker, a book on canal boat painting, and a series of Nicholson guides from the 1970s—which will serve us better than the canal companions we're using currently.

It was lovely, Polesworth Abbey, with its fourteenth-century abbey gateway and gardens. The tower houses eight bells, dating from 1664 to 2004, with this verse on the ringing-chamber wall, conjuring up past generations and ages:

> Who will divert themselves with ringing here
> Must nicely mind to ring with Hand and Ear.
> And if he gives his bell an overthrow
> Pay sixpence, a forfeit for doing so.
> He who in ringing wears Spurs, Gloves or Hat
> Pay sixpence as a forfeit for that.
> All persons that disturbance here create
> Forfeit one shilling towards the Ringers treat.
> Those that to our easy laws consent
> May join and ring with us, we are content.
> Now in love and unity join a pleasant peal to ring
> Heaven bless the church
> And George our gracious King. Amen.

Again and again, as we visit old churches, I have a sense of a past of meaningful continuity, and then the doubt sets in. These very continuities are under threat. We are becoming a people without a past. No longer do the narratives of history seem relevant, such is the enormity of the loss of the future. We have become unshackled from the past in other ways too; for our attention is all on the instant. My rebellion kicks in, and so I include doggerel like this. It conjures up the social relations of the bell-ringers of a past age, their humour and cohesion. Such things we must remember, or we unravel.

Bridgewater, Wedgwood, Brindley

Saturday began at Fradley Junction, a village with no great or deep roots, but created because of the canals, as the Trent and Mersey was joined to the Coventry in the late eighteenth century. It's a strange place. The canal companies built workshops, wharves, and warehouses here for the boats that carried coal, clay, and finished products from the Potteries, and beer from Burton-on-Trent.

The three key drivers were the Duke of Bridgewater, a major shareholder in the Trent and Mersey Canal, Josiah Wedgwood, who needed to get clay to his factory at Etruria in Stoke-on-Trent, and James Brindley, the imaginative genius of an engineer who built the infrastructure required. The Trent and Mersey was completed in 1777, joined by the Coventry Canal in 1789. There's a reservoir at Fradley too—an ingenious method to prevent water escaping into the Coventry Canal by taking the flow from above Middle Lock, around the Junction, under the Swan Inn to Fradley Pool, where it was stored until needed on the Trent and Mersey.

Fradley today is for the tourists. I wasn't sorry to leave. Off up the five locks that raised us ten metres along a half-mile stretch of the Trent and Mersey as we headed north-west. It's possible to do them all in seventy-five minutes, but it can take over three hours if there's lots of traffic. We turned left out of the Coventry, and were immediately into the top two, which went smoothly, without too much queuing.

Onwards towards Great Haywood, our destination for the night, through the most beautiful scenery, in glorious sunshine: Wood End

Lock, and then Handsacre and Armitage, and into Rugeley. There was another lock at Little Haywood where there was a queue of four boats already, taking almost an hour to get through, and then on into Great Haywood where we moored up, near the junction of the Trent and Mersey and the Staffordshire and Worcestershire Canals.

◆ ◆ ◆

We wandered to the pub, The Clifford Arms, had a pint and a meal, and reflected on how busy these waterways would have been back then: the enterprise, the urgency. A different world, where business interests started to shape society in new ways, driving change and disruption. Jenny and I explored towards Shugborough over the old packhorse bridge, the Essex Bridge, which spans the River Trent a hundred metres downstream of its watersmeet with the River Sow. It was built in the late sixteenth century by the Earl of Essex, who lived nearby at Chartley Castle, and is now the longest remaining packhorse bridge in England, with fourteen of its original round span arches left. It was old as the canals were built.

Shugborough is an incredible pile that we'd been able to see for miles from the canal. The present house dates from 1693, with the house and park improved by Thomas and George Anson in the 1720s–40s, then in 1762 with work by James "Athenian" Stuart, who built stone monuments called the "Tower of the Winds" and the "Lanthorn of Demosthenes". The Park Farm was designed by Samuel Wyatt towards the end of the eighteenth century. Illustrating just how the new industrial scene had its impact, the old village of Shugborough, in the grounds of the big house, was bought up and demolished by the Anson family, with villagers rehoused at Great Haywood. To give them more privacy and space, they moved a whole village!

Perhaps it was natural justice that the Anson family faced crippling death duties in the 1960s. The National Trust bought the place, leasing it to Staffordshire County Council who then managed the whole estate. Today it's back in the hands of the National Trust.[118] Again and again, the land around us held reminders of just how the rich have controlled and changed the lives and ways of life of countless ordinary people.

There's a lock down at Marple

For the last couple of weeks, we've seen the boat *Coningsby* and talked with her owners. It's what happens on the canals. You become part of a fluid community. They were there, moored in front of us, as we waited for the tree to be cleared; but further back than that, we first met them in Northampton, as Jenny and I set out. They overtook us at Fradley; we overtook them as they stopped to watch the Royal Wedding, and on Saturday evening they moored behind us at Great Haywood. *Coningsby* is borrowed from a friend as Mark is fitting out his own boat. I ask where, and he says, "Mirfield". Small world. He's a Huddersfield man, and knows the Leeds and Liverpool like the back of his hand. We meet his dog Skipper—a poodle/Jack Russell cross—and chat away.

Mark advises continuing up the Trent and Mersey (rather than taking the Macclesfield, as I'd planned) and coming into Manchester from the west to pick up the Leigh branch of the Leeds and Liverpool. Two reasons, he says: "There's a lock down at Marple. It should be mended by the end of the month, but it's been delayed already, so no guarantee you'll get through. Then, if you do, the 'Rochdale nine' are really tricky in the middle of Manchester. They have no holding pools, so the water flow is difficult to manage. They're underneath office blocks so can be claustrophobic—which might or might not bother you." "It bothers *me*," says Joan, his wife. Unlike others who have expressed an opinion about Manchester, he's not prejudiced against Canal Street, or concerned about security. Instead he commends the Bridgewater as stunning, and with fewer locks. So that's what we'll do.

◆ ◆ ◆

That evening I reflect on being a priest, prophet, pastor, preacher. Rubem Alves would add poet, warrior. We need poets and warriors as never before. That's what this voyage/pilgrimage is all about. Some of us are natural activists, and I applaud them. What I like about Jem Bendell's work is the way he roots his activism in a psychological understanding of grief, necessary in a world that is waking up to the loss it faces. Christianity offers traditions—as do all other religions worth the name—that hold together contemplation and action. What does it mean to be a warrior who is also a poet?

What more can I do to turn my future angst to the good? For the benefit of the creation which is all around us? There's a freedom, when you're past ambition. The world needs poet-warriors, with time to look above the parapet and wonder aloud how things might be different. There's a groundswell of cultural change happening, and Christian voices need to be part of that. Facing up to climate catastrophe is a moral issue—it concerns our mores, our way of life as part of Western civilization. It is a spiritual crisis as much as an economic and ecological one. It takes us to the heart of what it means to be a human being. At their best, human beings are contemplative, deeply reflective; able to find themselves in an engagement to that which is Other. Traditionally, and ultimately, that Other has been called God.

How can we commit ourselves—take vows—to change? To live more simply, deeply, slowly? With a sense of assurance that whatever happens, nothing can separate us from the love of God in Christ Jesus? I had grown so used to the habit of anxiety over the last few years that it felt good to sense a new resolve beginning to take shape. I found the words of Psalm 19 and read them, quietly, aloud; words that spoke of the covenant between God and God's creation. A covenant that will endure, despite the hard and difficult times ahead; a covenant that tells of history passing on its song, into the future:

> The heavens are telling the glory of God
> and the firmament proclaims his handiwork.
> One day pours out its song to another
> and one night unfolds knowledge to another.
> They have neither speech nor language
> and their voices are not heard,
> Yet their sound has gone out into all lands
> and their words to the ends of the world.
> In them has he set a tabernacle for the sun,
> that comes forth as a bridegroom out of his chamber
> and rejoices as a champion to run his course.
> It goes forth from the end of the heavens

and runs to the very end again,
and there is nothing hidden from its heat.

Psalm 19:1–6

Stone and Stoke

Sunday 21 May saw us moored up at Great Haywood. I set off for St Stephen's Church, which was nice and full for Pentecost Sunday. The preacher began her sermon with the disclaimer that she wasn't going to wave her arms about and go on for forty minutes. She, and everyone else, had seen the Rt Revd Michael Curry (the Presiding Bishop of ECUSA) the day before, stirring the congregation at the royal wedding with his powerful words on love and fire.

Jenny had checked the Australian coverage. Annabel Crabb of ABC News was impressed: Bishop Curry's sermon was different from the "standard Church of England sermon, which tradition dictates should be delivered in the tone of a very shy person asking the way to the train station".[119] The Word of God. Our words. What is preaching, in a world where what grabs and holds the attention today is big business? The attention economy is upon us, where subtle and not-so-subtle advertising intrudes, and it's hard to work out what to focus on, what deserves our deep attention.[120] For the Bishop, it was love.

I like Matthew Crawford's book *The World Beyond Your Head*.[121] He has written about attention and the attention economy, challenging me into a real engagement with the hard matter of things—like steering a narrowboat with skill. Digital companies now compete for our attention by their trade in clickbait. So how do we hold on to the discipline of giving deep attention to the natural world around us, instead of escaping into virtual reality, every moment of which is big money for one or other of the big global Internet companies? How do we attend to what makes us fearful, rather than distracting ourselves away from what really matters today? When we're living with deep anxiety, like all humans in their fear, we react by forgetting the grace of God, turning away into mind-numbing trivia. We stop looking at and caring for the natural world, which shouts aloud of God's presence through every being that has life. The world around us tells of God's glory, of God's faithfulness, of the living Christ, crucified and risen. The world, alive with the promise of God, yet troubled by a future that promises so much threat.

◆ ◆ ◆

As the time of Easter now turns with the coming of the Holy Spirit at Pentecost, it is always, traditionally, a time of blessedness. A time to think of all life pulsating with God's blessing. A time to remember that we live the resurrection life. There are reminders all around us—from the natural world, in the words we hear, the sounds and sights that speak of God's grace. Following Jesus Christ, I'd been baptized to live that life in all its abundance, called into greater, deeper, broader love. A Love that does not let us go; a love that breaks our heart, that we might know the abundant life of God. It's a time of the year when we move from darkness to delight. When in the silence and sound we can find sight and insight, to see the intensity of what lies before us.

So often, things are not known because they are not looked for. Or seen but not seen, because our eyes have become lazy and take the real, natural world for granted. Like light, for instance—the experience of light is common to us all; it is all around us. In the light of a grey day, a radiance can still shine through the golden-green new leaves and remind us of the light of another world. We can develop eyes to see the light that enlightens, the light that is immanent in the world, alive with colour, painting the meadows with delight, the trees and waysides with love.

That early summer it seemed as if we walked through a world enlightened by God's love and life. With eyes to see, there are bright fields that hold the presence of the one who is the light of the world. I tried to gaze right into the beauty, the gift of colour, of sight. I desired to be absorbed, to sink into the light, to become one with it so the light could shine into and through me, with a bright radiance that irradiated all the anxiety and fear. I sang the words in the light that "all shall be well, and all manner of thing shall be well". I wanted with all the yearning of my small heart to enter again the world of the child, to recover again the childlike lightness of being that brings us closer to eternity, the eternal now that is as present with us as the light surrounding us. I sat in the

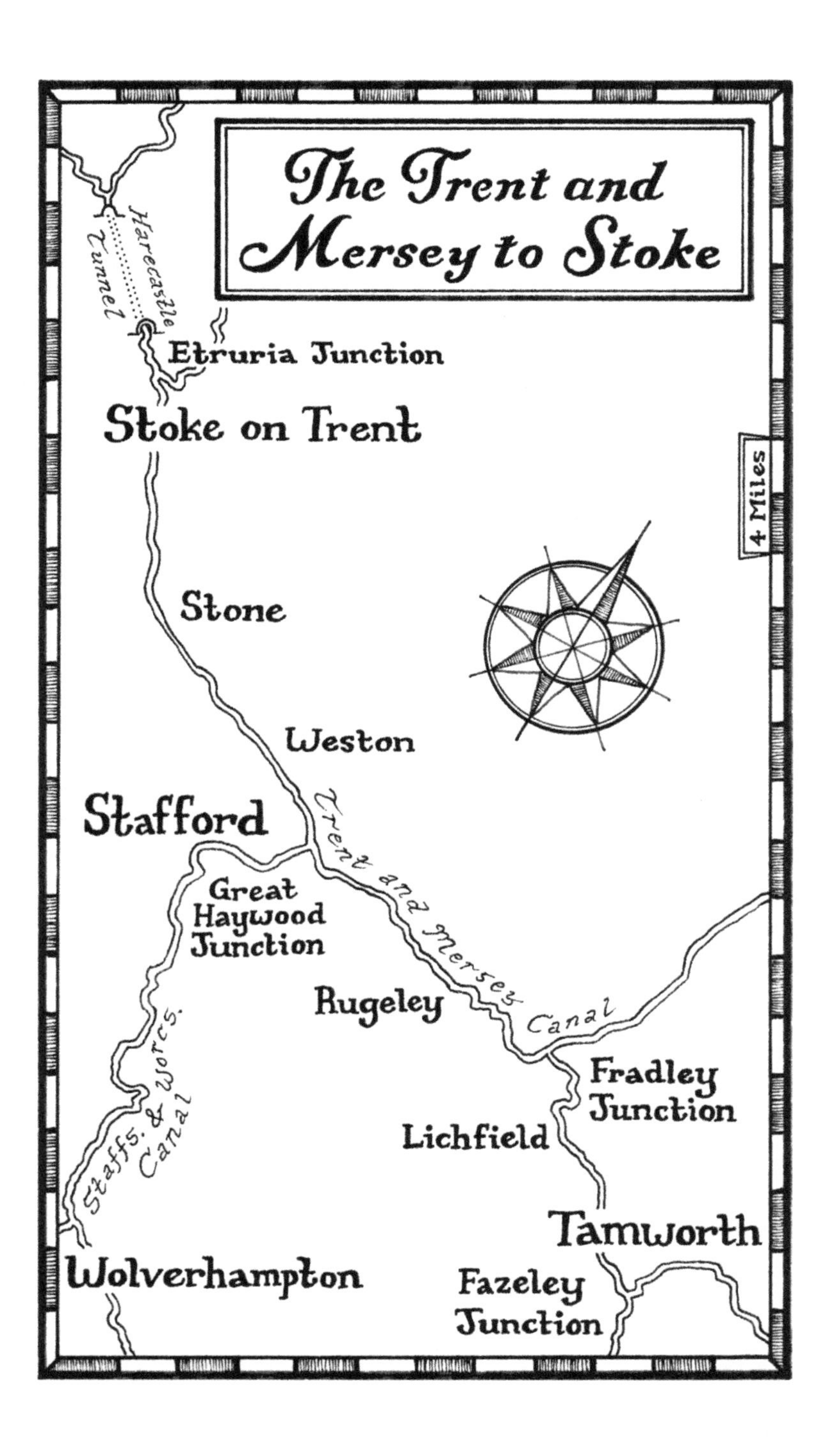
The Trent and Mersey to Stoke
Harecastle Tunnel
Etruria Junction
Stoke on Trent
4 Miles
Stone
Weston
Stafford
Trent and Mersey Canal
Great Haywood Junction
Rugeley
Staffs. & Worcs. Canal
Fradley Junction
Lichfield
Tamworth
Wolverhampton
Fazeley Junction

churchyard, surrounded by forget-me-nots and bees, and wished like a child, as hard as I could, eyes tight, turning my face into the full light of the sun. I wanted with all my being to wind the clock back to when I was eleven, in the 1970s, so that then the whole world, all humanity, had turned to the light and learned to love the natural world. To care for it; to work with God in the knowledge that all things belong within a unity made and sustained by God's love. I tried to attend to the light with everything I was, at that moment. Oh, the pain of it.

That morning, after church, Jenny and I walked back over the Essex Bridge to Shugborough and watched the light dance on the water as it flowed under that extraordinary monument. Light fractured and fragmented, carried away downstream.

The man who made water flow uphill

And then off we went, through Hoo Mill Lock, Weston, Sandon, Aston Locks, and into Stone, through Star and Yard Locks, to moor for the night right in the middle of Stoke-on-Trent. We enjoyed the hospitality of old friends who gave us gin and water (much to be recommended) and a fantastic meal.

This is still Brindley country. His workshop was opposite our mooring: derelict, though listed. By all accounts, Brindley was a brilliant civil engineer. Born in 1716, he worked his apprenticeship and made a name for himself designing and building systems to drain mines, restoring water pumps. He earned the name of "the man who made water flow uphill". So when the Duke of Bridgewater wanted a canal from Worsley to Manchester, Brindley was his man. With John Gilbert, Brindley turned his attention to extending westwards across Cheshire to the River Mersey, to Liverpool. Bridgewater had already purchased land for docks in Liverpool, so after the requisite Act of Parliament, Brindley started work on the Bridgewater Canal.

This we will join at Preston Brook. The distance of thirty-eight miles is posted on the towpath milestone.

With the Duke of Bridgewater's enthusiasm, Brindley's genius connected the Mersey to the Thames and eastwards to Hull, giving the Bridgewater

Canal access to the Potteries, and all the way to London and Bristol.

Josiah Wedgwood—another key player.

Wedgwood had worked up from humble beginnings. He'd suffered from smallpox as a young child which had left his right leg permanently weakened by an infection called Brodie's Abscess, leading to its amputation in 1768. That didn't stop him transforming the pottery industry. The only mode of transport was packhorse and wagon, which was slow, and costly in breakages. To create white pots required flint from South East England and china clay from Cornwall, so it was greatly to Wedgwood's benefit to see canals built that would enable access to his English customers, but also open up markets in America (through the ports of Liverpool and Bristol) and Europe.

The Trent and Mersey Canal Company held its first meeting on 30 December 1765. Brindley made such a good case that a bill was brought to Parliament, enabling the construction of the Grand Trunk Canal (as Brindley suggested it was called) to begin. Brindley predicted that the whole canal would be finished by 1772, but this was not to be, largely because of delays with the tunnel through Harecastle Hill, west of Stoke-on-Trent. It lay before us, to be travelled on Monday.

◆ ◆ ◆

A bloke on the tow path said to be there for ten o'clock, when the staff would check the headlight and the horn. We would need something warm for the forty-minute transition, a torch to see the roof in case of obstructions. We'd heard more than once of the death of a man, a few years back, who'd knocked himself out, and into the canal. The tunnel had been closed for days while police divers searched for and eventually found his body. There's a female ghost, apparently. And a skeleton painted on the right-hand wall.

There are now three tunnels through Harecastle—one built in 1827 by Thomas Telford, and one by British Rail in 1848. All suffered difficulty in construction. Brindley's tunnel was begun in July 1766. It's 2,880 yards long, nine feet wide, and twelve feet high. The first ever tunnel built solely for transport, it had to be dead level and straight. Brindley went through rather than around the hill, which would have required several locks and a water supply that wasn't to be had. A line was surveyed over the hill, with fifteen intermediate shafts dug down to the proposed level, and work was started at each end—so thirty-two faces were worked at once, cutting through a variety of difficult rock and quicksand. The stone was so hard it was almost impossible to penetrate—and the presence of water was a great drawback. Brindley died in 1772, so didn't see it out, leaving the work to be completed by his brother-in-law Hugh Henshall, in 1777. The tunnel had no towpath, so the boats were propelled by the boatmen legging it. Their reward, for almost one and three-quarter miles, was one shilling and six pence.

Stoke, Hanley, Burslem, Tunstall, Longton, Fenton

Monday saw us driving through Stoke, joined by old friends, Faith and Keith. Keith told us that it's really a federation of six towns, with Hanley the primary commercial centre. The other four towns are Burslem, Tunstall, Longton, and Fenton. Stoke, the centre of the Potteries: there is so much history to this fascinating town, yet so little money these days. We chugged through wastelands, where terraces had been demolished for new housing that was never built; we went silent in the shadow of derelict warehouses and kilns. We passed the branch of the Caldon Canal, where the Etruria Industrial Museum is situated. Middleport, where HRH Prince Charles has given support, is the world's oldest working Victorian pottery.

We rose through deep, deep locks that carved through the centre of the city. *The Lark Ascending*, twelve feet down below, seemed small as the massive gates closed behind and water poured and gushed in front, lifting us to ascend to the new ground level. It was all ominous, heavy. The town felt sad, so neglected. William Blake's words came to mind. I sang, against the roar of water:

> And did the Countenance Divine
> Shine forth upon our clouded hills?
> And was Jerusalem builded here
> Among these dark Satanic Mills?

The city's ceramics collection is housed in the Potteries Museum & Art Gallery in Hanley—and most of the major pottery companies have visitor centres, including the ten-million-pound Wedgwood Museum visitor centre opened in the firm's factory in Barlaston in October 2008. There are also smaller factory shops, such as Royal Stafford in Burslem, Moorcroft in Cobridge, and Emma Bridgewater in Hanley: I have a collection of her bird mugs, hanging up above the sink. They came from here.

◆ ◆ ◆

Keith says Stoke folk don't have much and don't expect much. There's a sense of betrayal: raised hopes of regeneration that never happened.

Working in the potteries would have been grim—lead poisoning a reality, causing early death and disability. From the late 1980s and 1990s, Stoke-on-Trent was hit hard by the general decline in the British manufacturing sector. Numerous factories, steelworks, collieries, and potteries were closed, and the sharp rise in unemployment was further compounded in 2008. It's a city crying out for a new start—like so many places in the North of England. It's thought to be one of the most cost-effective places to set up a new UK business, with affordable business property, surrounded by the Peak District National Park, close to Stone and south Cheshire, with excellent road links via the A500 and nearby M6 and rail links.

Will it happen?

The realities of the North of England began to sink in

The ravages of global capital care little for the social costs of unemployment. As this time progressed, my heart sank further and further as I contemplated the disaster that was Brexit. So many in the town, as in the North, will have voted to leave Europe. My only hope was that the costs would emerge as so immense that the people would see the sense of remaining after all. Then, within Europe, we can all begin to tackle the challenges of climate change in earnest, which will, I'm afraid, mean the poverty of cities like Stoke will stack up even more, leaving the left-behind even further behind.

What of the lifestyle changes that we will all have to make? It's much easier if you can afford them. What if you're living in a town like Stoke, or Wigan, or Burnley, where there's no give for anything—where Universal Credit leaves people struggling for food, for decent clothes? Where a holiday is a dream?

Living a low-carbon life . . . many are beginning to embrace it. They don't think we can wait until 2050—and we can't—to cut greenhouse gas emissions to zero.[122] This will mean the end of petrol and diesel cars (and narrowboats?). The end of gas boilers. So people are already developing eco-superhomes, which have been adapted to achieve a sixty per cent or more reduction in carbon usage. To make a home energy efficient, it needs insulation—floorboards, ceiling spaces, walls; double glazing, solar panels on the roof, and a battery bank to store the energy. Many are already embracing diets that are vegan and vegetarian, as meat production is not ecologically sound. Increasing numbers grow more fruit and vegetables and eat seasonally. We should start mending our clothes and furniture, shopping in charity shops, using sites like Freecycle[123] or Freegle[124] to advertise stuff we no longer need. There are banks, like Triodos,[125] where money is used for ethical and sustainable purposes, not fossil fuel exploration. Some people convert chest freezers into fridges—so the cold air doesn't fall out—using half the electricity. There are electric cars charged with solar panels; even cooking outside in sun ovens, in hayboxes. There's wastewater from the sink to use in the toilet. Air source heat pumps. Recycling and repairing rather than replacing. All this is already happening—but it takes resources and the luxury of time to plan and implement. Resources that only those who are affluent can afford.

This is relinquishment, and restoration, as Jem Bendell advocates, against the time in the future when everyone must live with the unpredictability of climate chaos. But what if you haven't any resources to make these adaptations now? What if it's hard enough just to survive, to eat? Where is the political will and action that is putting proper funding into places like Stoke so people can begin to live with hope? Where is the deep economic framing that moves away from the stranglehold of neo-liberalism, with its mantras of hyperindividualism, atomism, incrementalism, all in the mistaken belief that the market will turn up trumps? The thinking is there already—it's been around since Schumacher published *Small is Beautiful* in 1973: a study of "economics as if people mattered". There are many who have continued to espouse and develop such ideas. The people do matter. We face the scenario graphically described in Marge Piercy's book *Body of Glass*, where the rich live in great Eden-Project pods, and the poor in uncivilization and utter lawlessness. The polarization between rich and poor is steadily widening.

Now is the time for change.

The psalms are full of verses that speak out God's justice for the poor of the world—who will be the least able to adapt to climate change; who will suffer the ravages of the undoing of civilization as we know it. It is not God's will that the richer are richer, where wealth now is cynically gained through distraction, clickbaiting the attention of the hopeless and helpless so the poorer become more impoverished—materially, socially, psychologically. Now is the time to invest in towns like Stoke, Hanley, Burslem, Tunstall, Longton, and Fenton. To encourage local and regional economies with new ways to adapt in order to become resilient, to relinquish what's not essential, to restore what we already have.

Psalm 102 captures where God's will can be found—even when our lives are transient, God's justice cries out for the destitute, for the cities of the earth. I made it my cry:

> My days fade away like a shadow,
> and I am withered like grass.
> But you, O Lord, shall endure for ever
> and your name through all generations.
> You will arise and have pity on Zion;
> it is time to have mercy upon her;
> surely the time has come.
> For your servants love her very stones
> and feel compassion for her dust.
> Then shall the nations fear your name, O Lord,
> and all the kings of the earth your glory,
> When the Lord has built up Zion

and shown himself in glory;
When he has turned to the prayer of the destitute
and has not despised their plea.
This shall be written for those that come after,
and a people yet unborn shall praise the Lord.
For he has looked down from his holy height;
from the heavens he beheld the earth,
That he might hear the sighings of the prisoner
and set free those condemned to die;
That the name of the Lord may be proclaimed in Zion
and his praises in Jerusalem,
When peoples are gathered together
and kingdoms also, to serve the Lord.

Psalm 102:12–23

The light was a pinprick—forty minutes in front

The 22 and 23 of May saw us leave Stoke-on-Trent, on our way to Sandbach, and from there to Northwich. But Northwich is arrival. Before that, many locks—many, many locks—down what's called Heartbreak Hill onto the Cheshire plain, and before that, the incredible Harecastle Tunnel. We were given our safety talk; warned that some of the sections go as low as five feet nine inches, so to mind our heads. A torch at hand is useful. That it gets cold and wet, so to have waterproofs. To travel forwards, always, and not to reverse. To leave a safe distance between you and the boat in front. We were told that if we didn't come out of the tunnel in an hour and a quarter, they would monitor our progress and decide whether (or not) to call the emergency services.

We had to show our headlight was working and sound our horn. This was a moment of anxiety, for the horn hadn't worked for ages. I'd been in touch with Alan, at Fox's, but he couldn't suggest anything apart from fiddling with it, which didn't work. Faith had the great idea of using an old, long, brass hunting horn they had at home. So we had that with us, in the hope that we could convince the tunnel staff it would work instead. Then, just moments before the tunnel keeper arrived, I gave it

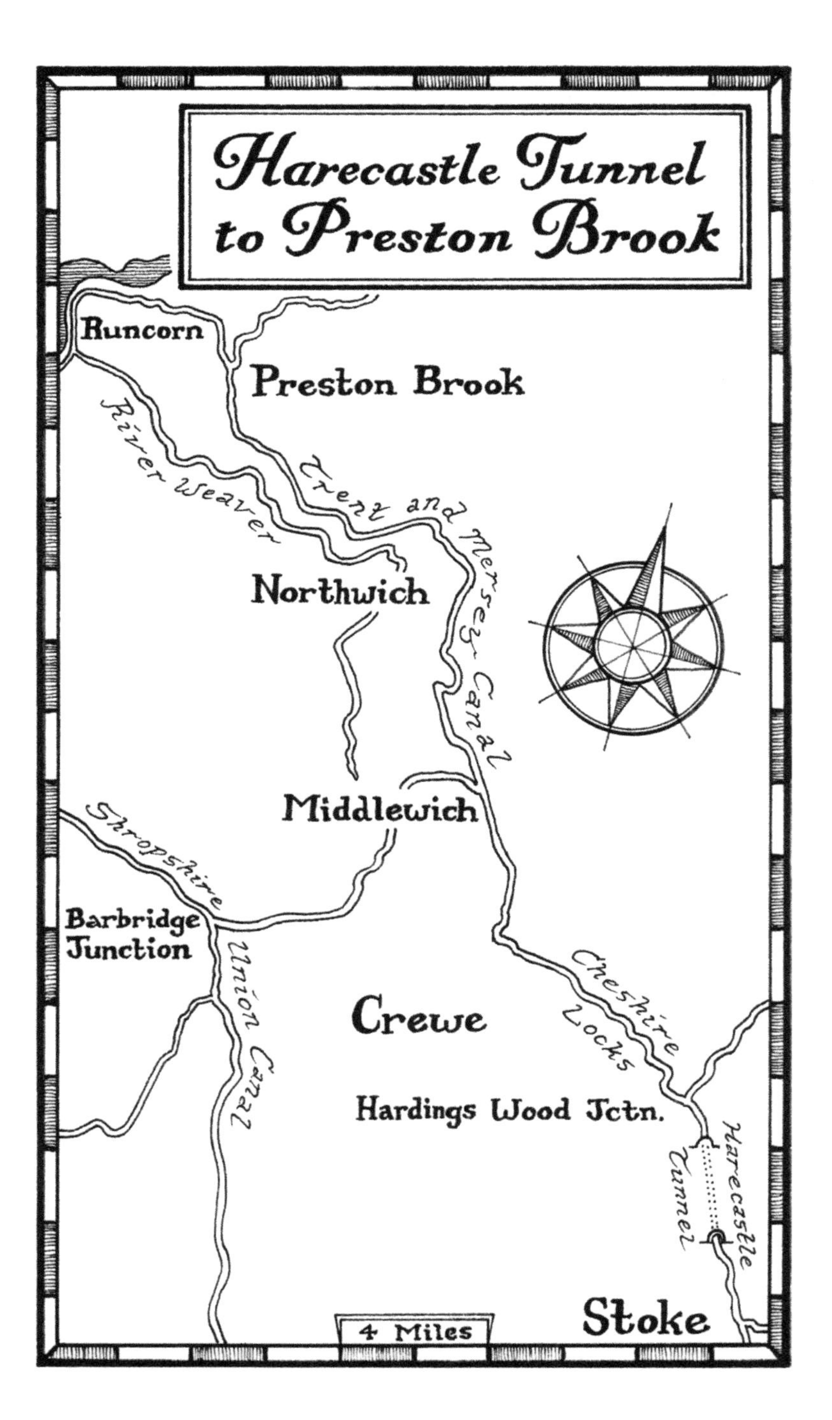
Harecastle Tunnel to Preston Brook
Runcorn
Preston Brook
River Weaver
Trent and Mersey Canal
Northwich
Middlewich
Shropshire
Barbridge Junction
Union Canal
Crewe
Cheshire Locks
Hardings Wood Jctn.
Harecastle Tunnel
Stoke
4 Miles

another go, and it blared out. A loose connection somewhere; a relief, as I didn't think he would be impressed by Keith and Faith's alternative. If in trouble, we were told to give one long blast every thirty seconds and repeat until three short blasts came back. The hunting horn would have done the trick—but I was glad not to argue it out.

The tunnel is nearly two miles long. It takes forty-five minutes of chugging in the dark. All the way you can see the light at the other end. "That's incredible!" said Jenny. Looking back, there was only intense, black darkness. Imagine—well, I did—legging it through as men used to do, to the sound of silence made louder by the dripping water, the periodic gushing water feeds. Did they shout to each other? Keep up banter, or sing all the way through? Or was it just long, silent slog?

Incredible—to build so straight over such a distance. Brindley dropped fifteen shafts, which then were worked in both directions from each, and also from each end. How it all met up is beyond me. And the light was a pinprick—forty minutes in front.

Faith, Keith, and Jenny spent most of the time in the fore cabin, watching the ongoing darkness, drips, and brickwork, leaving me to concentrate. I needed to, to ensure the boat stayed in the middle. Once or twice we hit the sides and lost wood from on top. The trick is to keep the arc of light from the headlamp as equal around the boat as possible, and to focus on the light at the end of the tunnel. It was hard, though. If you got too close to either side, there was little leverage to regain the centre. The light at the end of the tunnel—never has a metaphor had so literal a meaning. In the depths of dark despair, I pray for light. Even such as this—so far ahead.

It's in those bleak, liminal times before dawn that the eco-anxiety hits hardest. I lie in bed praying, desiring with all my heart that there be a future; that I might have again that taken-for-granted confidence that I'll live to see grandchildren, great-grandchildren. Now young people are deciding not to have children at all.

It's holding together the level at which reality poleaxes you with fear and anxiety, and the "putting one foot in front of the other" level of daily life. Really hard to cook your next meal, if you've decided not to have children because you don't think there will be a viable future. Someone once said that lament is the antidote to despair. And in lament is desire, deep desire for things to be different. Harecastle Tunnel was a weird mixture of intense concentration on steering as best I could, and a deep desire for a light in case the darkness lasts always. Almighty God to whom darkness and light are both alike, to whom all hearts are open, all desires known. It's not good to be left alone with such thoughts; (or is it?).

Heartbreak Hill

As we came out of Harecastle, it was wonderful to have Keith and Faith with us. It gave Jenny a well-earned day of rest, as we descended over twenty locks to the Cheshire Plain, colloquially called "Heartbreak Hill". We talked and walked; struggled with ratchets and heavy gates; enjoyed omelettes, and flapjack made by Faith, and reminisced of many things. Keith even slept on the towpath. Friendship is very special. We haven't really seen each other for thirty years, drifting apart through the busyness of life. But it's as if nothing has changed. Except we're all a bit older.

They left us at Sandbach, walking to the station to catch a train to Crewe, and then back to Stoke. "Great not to have to do all those locks with only me," said Jenny. We moored up for the night, and again I was thrashed at Scrabble.

Next morning I set off for some milk.

You've got a lot of stories in you

This silver-haired man stood behind me, looking quizzical. He asked my name and repeated it after me. "Frankie. Frankie." Seemed he liked it a lot.

"That accent's not from here." "Guess," I said. "Further south," he hazarded, and so I explained how I was travelling with an Australian, and some of it had rubbed off.

I asked him if he'd always lived here, where I'd called in on spec to have my hair cut. His mother came from Cork, and married his father, who was a butcher from Wilmslow. His father recently died, a real character. He had loved his mother—there she was, a photo in the corner.

He told me he liked the colour of my hair. "It's the same as yours," I replied.

"But yours feels better."

He looked quizzical again, flirty, saying nothing, but obviously waiting for me to tell him what I wanted. "As I walked along, I thought I'd like a number twelve. Short," I said. He started cutting. "So?" was in the air. I didn't give much away, enjoying this man and the games he played. "You've got a lot of stories in you," he said.

"Travelling." He had picked that up and presented it, a statement rather than a question. "I'm taking a narrowboat from March to Skipton. I'm on the canal."

"Oh, no. No. No. I wouldn't like that."

He told me he'd owned the business for twenty years. He was settled. "I don't have the same drive I did. You leave it behind when you're our age. Other things become important," he said. I was impressed. Some hairdressers have it. The ability to probe with real insight into their customers' lives. I intrigued him. I could see that. The pitch of his ear for others was acute.

"So why?"

I explained we were relocating from Suffolk to Workington in Cumbria. That the boat would be kept in Skipton. I knew he wanted to know why. He said it again. "You've got stories in you."

I said, "You're good at listening." He described how sometimes he just wanted someone to shut up. They would go on and on.

Silence for a while. Our eyes caught—both of us flirtatious. I gave in. "I'm a vicar. I've been running a busy cathedral. My husband's training to be a vicar. I've just finished a book. I want to write more," I said. He enjoyed his own sense of surprise.

"So you believe there's something when we've gone?"

"Of course." I launched in. I don't do wishy-washy liberal anymore—or maybe I do. "I think love is the most important thing, and when we die, we are taken into God's love, into God's everlasting arms. Don't know what it means beyond that; don't think we can know." He told me of a funeral he'd gone to in Ireland of a friend of his aged fifty, with two small children. The priest had begun the service by saying, "God is good." He said he couldn't believe it: that was the last thing the family and gathered mourners wanted to hear. God wasn't being good to them. "But that's Irish priests for you."

He changed the subject by talking to his colleague about a customer, and then whether hair trimmers were made in size twelve. She said yes, and went to look.

"Were you christened, Frankie?" "Of course," I said. "No. Were you christened 'Frankie'?" "Frances," I said. "I became Frankie when Frankie Goes To Hollywood were in the charts."

"Thought you were that, my sort of age."

"1959," I answered his question. He told me he was sixty-two.

"Sixty is the new forties," I opined. "It's a great decade. I plan to enjoy every minute." He looked like he was going to as well. "It's taken five years off you," he said. I grinned. "I don't like to look mumsy," I said. "You look like a writer now," he said.

"Do you have any bad habits?" "None I'm telling you," I said. I was enjoying the me who was enjoying the flirting. I looked at myself in the mirror. Impish, cheeky, my eyes on fire.

I got up to pay. Twenty quid, cash. He was onto the next customer. As engrossed with her as he had been with me. A priest in his own way.

Salt of the earth

Back on the boat, with eight more locks ahead to bring us down onto the Cheshire Plain, we left Sandbach in the glorious sunshine we now expected. Jenny had brought it with her, of course.

A lovely day, along a stretch of canal between Middlewich and Northwich. It's green and tree-lined as it contours around the valley of the River Dane. We curved around towards the east, heading for the town of Northwich. It was quiet and beautiful.

Any town that has "wich" in it—Droitwich, Middlewich, Northwich, Nantwich—will be a place where salt has been made, probably since Saxon times, for "wich" is a Saxon word. We passed great chemical factories and mountains of salt.

The canal has subsided again and again through this section, because of the mining. We pulled up outside the Lion Salt Works at Marston:

too late, unfortunately, to see around the works, but not too late to buy a guide, and have a pint in the Salt Barge pub, where all over the walls are pictures of massive salt mines.

I had no idea, when living only a little way north of here, that this was such a salt-producing area. According to the guidebook, as Jenny read it out over our pints, the geology of Cheshire is a large shallow basin formed between the sandstone ridge of the Delamere Forest to the west and the Cheshire hills to the east. In the Triassic Period, 220 million years ago, this was a large tropical lagoon that trapped sea water. Evaporated, it became halite, better known as rock salt.

There are two bands of salt beneath this ground, each about twenty-five metres deep, separated by ten metres of brown marlstone. When ground water flows over the salt layers it dissolves the salt and creates underground streams of salty water or brine. The underwater brine was known as "Roaring Meg", when the pressure forced the water through cracks to emerge as streams.

There is evidence of salt production from before Roman times. Surviving fragments of clay pots show that the brine was heated until the water evaporated, leaving the hard salt deposit. Outside the Lion Salt Works is a large pan that did the same. With fires underneath, the workers would rake up the salt, skim it, and lump it in tubs, turning it out in rows.

This method requires fuel—coal. Mark Kurlansky, in his 2002 book *Salt*, tells how, in 1670, John Jackson prospected for coal on the estate of William Marbury, near Northwich. At a depth of only 105 feet, he found a bed of solid rock salt and no coal at all. Marbury was disappointed: he wanted coal. He went bankrupt in 1690. Only a few years later, Sir Thomas Warburton opened four salt mines in Cheshire, and made a fortune.[126]

I studied *Anthony and Cleopatra* for A-level. Pompey's description of the Egyptian queen has always stuck with me. "But all the charms of love, Salt Cleopatra, soften thy wanned lip!" Salt has always been associated with fertility, so brides would have salt in their pockets, or have their feet sprinkled with salt. We get the connection when we call someone salacious. Folk from Cheshire don't strike me as salacious. But perhaps I'm wrong. I've never had such a salty haircut.

The thought of those deep salt mines below stayed with me, were caught up in my dreams that night. The necessity of salt to life; yet the sterility of too much. Salt Cleopatra, whose desires ran away with her. It's the managing of emotion that's so hard. Not allowing myself to be overwhelmed by despair and anxiety; finding hope in the everyday routine things. Not letting the anxiety undermine to such an extent that there seems no point to life; not letting the fear grow so big that there's no room any more for God. Instead, trusting in the grace of God, who is the fulfilment of all desire, and allowing God to guide the discernment of which desires to follow and turn to action.

Lamenting with a fierce hope; then deciding how best to live the rest of my life. I heard on the radio of a teacher who had given up teaching, because he could no longer educate children for a future that was described only in terms of a particular sort of success, for it no longer rang true. So he joined Extinction Rebellion and started protesting. It felt good to him. It felt like *doing* something, rather than succumbing to helplessness.

What does that mean for me now? How do I work out what best to do, given my gifts and talents, to respond to God's grace with all the desire, the saltiness, I can bring? Planting apple trees. Digging a wildlife pond. Protesting at the lack of political commitment. Anything to answer those deep, dark fears of three in the morning. Anything to glimpse a light at the end of the tunnel.

CHAPTER 10

Bridgewater to Wigan Pier

Three tunnels—Barnton, Saltersford, and Preston Brook—mark the end of the Trent and Mersey Canal, and then there's a stop lock at the beginning of the Bridgewater.

The beauty of may trees in blossom and of wildflowers continues, including a bed of orchids just before the stop lock. There's ragged robin, columbine, campion, buttercups galore, archangel, forget-me-not, Queen Anne's lace, greater stitchwort, and speedwell, reflecting the sky.

I think of *The World My Wilderness* by Rose Macaulay, set in London after the Second World War.[127] How she tells the story of Barbary, the daughter of a voluptuous, indolent, but intelligent mother, divorced from her father, a successful barrister. She has been brought up in France in the immediate aftermath of the war, indulged, neglected by her mother, and sent back to London to be civilized by her father and his new family. She misses her childhood and her new young brother, and resents her father's new wife and child. She makes friends with Raoul, her cousin, and together they run wild in bombed-out London, looking for somewhere else to live.

Richard Mabey, in his book *Weeds*, describes Macaulay's London in the immediate aftermath of the war.[128] How the weeds came first: the flowers and colour into every nook and cranny, a renaissance of life and hope amidst the destruction and dark desolation of shattered lives and cities. Barbary is wise beyond her years. She has seen adult emotion, the break-up of her family. She has been shifted here and there and is homeless. Her life is mirrored in the state of London, and in her life the weeds begin to grow.

Where do we find signs of hope and life in our lives? Where are the weeds? Are weeds simply flowers in the wrong places, or are they

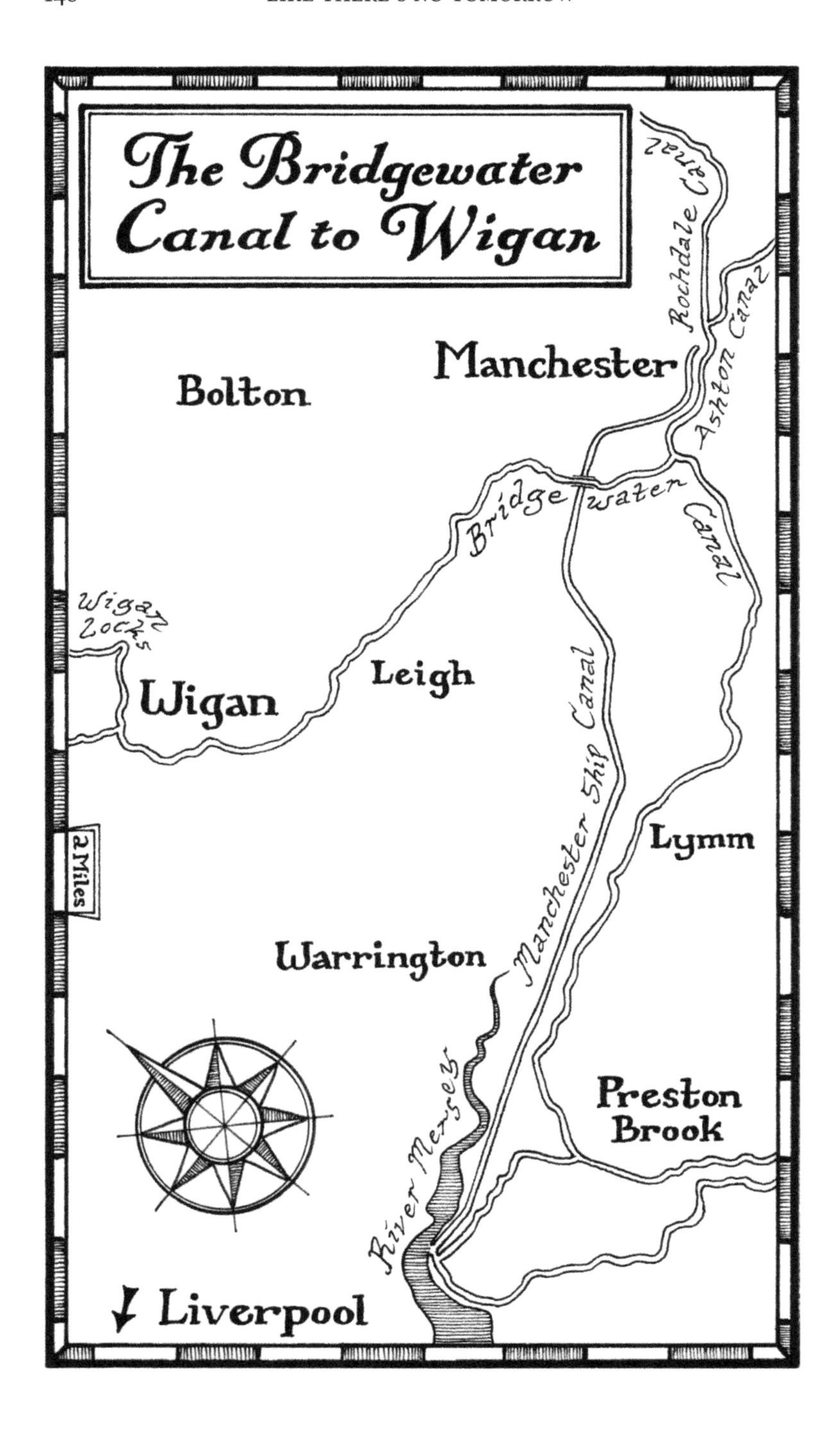
The Bridgewater Canal to Wigan
Rochdale Canal
Ashton Canal
Manchester
Bolton
Bridgewater Canal
Wigan Locks
Wigan
Leigh
Manchester Ship Canal
Lymm
2 Miles
Warrington
River Mersey
Preston Brook
Liverpool

a reminder of a more elemental life that is beyond our control, which might bring chaos, but also its own beauty?

Barbary and Raoul found a world:

> of little streets threading through the wilderness, the broken walls, the great pits with their dense forests of bracken and bramble, golden ragwort and coltsfoot, fennel and foxglove and vetch, all the wild rambling shrubs that spring from ruin, the vaults and cellars and deep caves, the wrecked guild halls that had belonged to saddlers, merchant tailors, haberdashers, wax chandlers, barbers, brewers, coopers and coachmakers, all the ancient city fraternities.[129]

It's a world that gives an answer (of sorts) to the emotional chaos of their post-war lives. And perhaps the weeds are the best place to go when our lives, or the lives of children we know, become chaotic. There's plenty of research suggesting that the best thing we can offer to a child we know who is unhappy, or anxious, is a nature walk. To put to one side, for the time being, all the unhappiness and anxiety, and learn, instead, to tell the difference between the weeds, the flowers, the wayside plants observed as we walk along. To see how many species can be named. To work out the difference between purple loosestrife and rosebay willowherb.

I remember, when we lived in Bolton, two ten-year-old boys who caused trouble in the neighbourhood. One evening, there they were, as I took the dog out, looking as if they were about to be up to no good. I knew them from assemblies in school—well enough to ask them to come with me to walk the dog. They almost didn't, awkward and embarrassed. But hearing the sound of a little owl, and my answering call with cupped hands, they were hooked. Before long, from that and other occasions, they knew hawthorn from blackthorn, elder from bramble, Spanish bluebell from English. They had seen a kingfisher. I wonder if their appreciation grew into a better stewardship of our natural capital.

It is so easy to be distracted from nature all around us: not to engage, as we live our busy lives, caught up in patterns of behaviour that commodify every element of existence. So easy to forget the free gift of the natural world around us, waiting to delight. I remember a car journey when there were red kites, soaring above. I tried to capture it in a sonnet, "Red Kite":

> Travelling north by car, all the while,
> red kites. They twist around and play the air,
> each tail feather tuned to response, as mile
> on mile we drive. Their skill I watch, aware
> the careful flight is geared to scan the ground
> for prey. It seems that all the kites are one;
> one following me, watching all around
> as counties, cities, fall behind. It's gone;
> and anxiously I search the empty skies.
> We are too used to being watched. I miss
> surveillance, strangely; in it some comfort lies.
> There! There! A sudden joyfulness
> I fling at the hungry sky. The silent kite
> stoop-dives its prey, and I am taken in flight.

◆ ◆ ◆

I went to help Jenny with one of the gates at the stop lock, and as I reached to step back on board, slipped and ended in the brink. Hey ho. No harm done, except to my phone, there in my back pocket after taking a picture of the last Trent and Mersey milestone, with Preston Brook one mile to go.

These milestones have been with us all the way along the Trent and Mersey. Shardlow one way, Preston Brook the other. Sometimes they have added up to the same amount of miles, sometimes not. It should be ninety-three and a half. Every so often the distances seemed to suggest some magical addition or subtraction.

But never mind that. My phone! Straight into rice, but it still wasn't working by bedtime. Or the next morning.

◆ ◆ ◆

We entered the Bridgewater, and Jenny began to sing "No more locks 'til Wigan!" as we began the long circle inland to Manchester, running alongside the great Ship Canal. It would take us out west again, through old stamping ground of mine, for I had served my curacy in Westhoughton, Lancashire.

As we chugged along towards Manchester, I persuaded Peter to leave his essay on sacramental theology (no easy task) and join us on Friday, so he would be with Jenny and me as we did our last day together. For Jenny was to leave for London and Morocco the following Monday, and I was to return to Mirfield for the inside of the week, until the end of term.

We planned to leave the boat at Wigan during the end of term, and move our stuff from Mirfield to Workington. Then Peter could have a well-earned holiday as we did the final leg together, up the twenty-one locks out of Wigan, and onwards towards the Pennines. Once the boat was in Skipton, we could move our stuff properly into the rectory at Workington.

As we have come north, slowly and surely, it's felt good.

Lancashire, then Cumbria, beckon

We're still in Cheshire, though, as we pass Warrington, and draw near to Lymm where we moor up to stay the night. It's a pretty town, with a village centre. The pub is just above our mooring, and we hear laughter and good humour as we eat a pie: broccoli, cheese, and lentils with tomatoes.

My old friend Harry used to quiz, "What's a balanced diet in Wigan?"

"A pie in each hand."

Early on Friday morning we set off, travelling through wonderful country, past Dunham Massey Hall (obscured by trees), then the outskirts of Greater Manchester, and into Sale.

Peter said he'd be at Victoria Station at two o'clock, so we moored up on the towpath at Sale Metrolink, and headed into the city. The first destination: an EE shop, so I could replace my waterlogged handset, which hasn't recovered despite that night in rice. Emma served us. She had a degree in Philosophy from Manchester University, and was returning there to do a PGCE in the autumn. She came from Wigan—and from a family of teachers. Now living in Didsbury, she was rather scathing about her birth town. I wanted my friend Eth there to defend it. The shop didn't have WiFi—it was down—so she advised us to go to the Apple Store in the Arndale Centre to update the settings.

There we talked with Lars (well, let's call him that). He was half-Swedish, half-American, and had done ten years in the US army. He reeled off all the places—ones you'd predict—he'd seen service. He told us he thought the UK was full of rage, glad he was only staying six months before heading back to Sweden.

He declared he was a Hell's Angel. I guessed they'd been going since the sixties, but it was more than that: "We're sixty-two years old," he said. He described the brotherhood, glowed about his Harley Davidson, and showed us his tattoos. I wondered what initiation rites there were, what they did together. "Let's just say, 'things,'" he replied. He commented that the law was much stricter here in the UK than in the US. He told us that a lot of the ex-military become mercenaries. "What, like vigilante?" "Yeah, that's right. We take out paedophiles, that sort of thing."

He was disarming, looking you straight in the eye. My phone finished uploading and I headed off to see if I could sell the old handset, leaving Jenny with her downloads and updates. She found out that he was adopted. She thought he was trustworthy. I wasn't so sure. I wish I'd asked him the last time he sang out aloud; the last time he cried.

We wandered into St Ann's Church, which was as beautiful as ever, and then the Cathedral, which is a transformed space since I was ordained deacon there on Advent Sunday in 1989: now with a new marble floor and a lightness and colour that lift the building from the heavy, dark interior that used to pull the space downwards and inwards, leaving little air to breathe. A stunning new organ fills the nave as if with light.

Manchester Cathedral

Shekinah. A cloud of glory, from which comes glorious sound.

I reflected that it's always been the case, that God has been there, my light and salvation, even through the shadowlands of anxiety and depression:

> The Lord is my light and my salvation;
> whom then shall I fear?
> The Lord is the strength of my life;
> of whom then shall I be afraid?
>
> *Psalm 27:1*

There was an interfaith day happening while we were there, and so I took time to visit the Fraser Chapel to see Mark Cazalet's triptych, so controversial when installed in 2002.[130] It's an image of the Blessed Virgin Mary with the two patron saints of the Cathedral—on the left, St George, as a young black man in jeans and trainers with a dragon on a string, and on the right, St Denys, a monk, holding his head in his hands as he walks through the shopping centre. The picture is shaped like a dressing-table mirror. The central panel shows a young man and woman eating fish and chips, with an old man in the centre, facing the viewer. There's a bottle of ketchup on the table. This is the Trinity—Father, Son, and Holy Ghost, eating fish, as Jesus did with his disciples after the resurrection. It's good to see a woman represented in the Godhead.

I enjoy my favourite window too—flaming, fiery red, installed after the bomb damage of the Second World War. That day, you couldn't help but think of other bombs. It was the first anniversary of the Arena bombing.[131] Manchester Arena is only a stone's throw away, and the Cathedral was there as the city remembered, with trees all around covered with prayer requests and remembrances. Back in 1996, we heard that other bomb as I watched the children at a swimming lesson in Bury, ten or so miles to the north, and felt the impact. It had devastated Manchester, but miraculously, no loss of life.[132]

Hate feeds off carrion, like despair. It feeds off resentment and anger, and wants to see itself reflected in its impact, in the hateful face that returns its gaze. It's so hard to forgive when someone deliberately seeks to do you harm, to kill or maim someone you love. That's where the psalms are so good. They take you right to the uttermost pit of hatred, yet enable you to find God there. Then hatred is not the last word. It can—will—always be overcome with love. Love stretches further, beyond hatred. Without forgiveness there is no way to go, except into the depths of hell. Even if your sentiment is this:

> Let them vanish like water that runs away;
> let them wither like trodden grass.
> Let them be as the slimy track of the snail,
> like the untimely birth that never sees the sun.
>
> *Psalm 58:7–8*

Then we can still say:

> God is our refuge and strength,
> a very present help in trouble;
> Therefore we will not fear, though the earth be moved,
> and though the mountains tremble in the heart of the sea;
> Though the waters rage and swell,
> and though the mountains quake at the towering seas.
> There is a river whose streams make glad the city of God,
> the holy place of the dwelling of the Most High.
>
> *Psalm 46:1–4*

The bombs can be metaphorical too—bombshells dropped into families: news of pain and anguish that is hard, so hard to bear; knowledge of the actions of others that wilfully seek to undermine the vulnerable, where you can do little to heal the situation, or to make it better for someone you love.

Jenny and I headed for Victoria Station to meet Peter. He didn't want to hang about, so we headed for the boat and chugged off towards Trafford Park and through the Barton Swing Aqueduct that took us over the Ship Canal—water on water, many feet below. Magnificent. At Parrin Lane Bridge there was a lighthouse, and shortly after, a sculptured reminder of the horses that worked these routes.

After Worsley

With St Mark's Church spire—that local landmark—just there above us, parkland extended on both sides of the canal. We passed through Boothstown and moored up just past Astley Bridge. It's a little-used, little-moored stretch of canal; the fewer boats, the less secure it seems.

It rained, off and on, all day. The first day of rain Jenny has seen. Though once we had the wood burner hot, and the bean stew eaten, the sky began to clear as it became darker. The birds sang loud and late. I read aloud a poem written to capture something of the resurrected life as Stanley Spencer saw it—"Nightingales at Cookham":

> Or shall we go by heaven to heaven
> as mortal minds and bodies toughen
> with joy, relinquishing the pain
> of loss? For then we shall be one
> as the river flows into the sea.
> We hungry two, who feed on bread and wine,
> consume that joy again.
> We stand in Cookham, stretching limbs, now free
> to love; to kiss, to eat and drink, and smile.
> The nightingales sing; you whistle, mile on mile;
> a world restored, my love, and all worthwhile.

The Saturday morning of the late May bank holiday weekend, with Peter and Jenny on board, we left Astley Bridge; three quarters of an hour later we were in Leigh. Our very old friends Harry and Eth joined us here. They come from Hart Common, a small village near Westhoughton, and have been friends since I served my curacy there in the late eighties.

We approached Wigan past old mine workings now filled with water—called, locally, "flashes". Pennington Flash, Scotsman's Flash—"Who was the Scotsman?" Jenny asked Eth. She didn't know. "Why's it called a 'flash'?" None of us knew that either.

Why's it called a "flash"?

A bit of research reveals later that a Flash is a stretch of water caused by the flooding of old mine shafts and pits with water from the canal. Scotsman's is now a site of special scientific interest—and the whole area of the Pennington Flashes, for there are a number around here, are valued for their wetland wildlife and biological interest. It's called "Scotsman's

Flash" because in the late nineteenth century this was where the Scots of Wigan came to curl. The Wigan and Haigh Curling Club was founded in 1861, and at its height had 168 members on roll.[133]

The way this whole area—where once the canal would have seen boat after boat, laden with coal, with filthy air and hard local lives—is now an extensive nature reserve gives hope. Like the weeds of post-war London. The waste land is where the water comes, filling those mine shafts deep underground where men once worked, and providing extensive stretches where the land has subsided, in places twenty feet below the canal. Land no longer useful, so nature has a chance again. Waste land. The Fens came to mind, where water will, one day, recapture and regenerate the wild, when humans have finished extracting what they can.

> O God, you are my God; eagerly I seek you;
> my soul is athirst for you.
> My flesh also faints for you,
> as in a dry and thirsty land where there is no water.
>
> *Psalm 63:1–2*

That ancient story of flood which is there in Genesis, and perhaps has basis in the formation of the Black Sea nearly 8,000 years ago, stirs primordial fear but signifies new beginnings too. The deep waters of death bring life.

The Leeds and Liverpool Canal

At Leigh Bridge, the Bridgewater becomes the Leeds and Liverpool Canal. We stopped the traffic by lifting the road to mark the place. The Leeds and Liverpool heads west to Liverpool at Wigan, or east towards Leeds up the Wigan flight—twenty-one big and heavy locks. Jenny breathes a sigh of relief that these won't be down to her. We planned to leave the boat for a week in Wigan, so we pulled up. Harry and Eth headed off, thrilled to have seen familiar territory from a different perspective. Peter went too, back to Mirfield. I'll join him on Tuesday. I bought a handcuff key (an anti-vandal key) from the Canal and River Trust man who happened to be passing, and Jen and I settled for the night, surrounded by apartments, with Wigan Pier just around the corner, and Trencherfield Mill looming above us. Built in 1907, it's formidable—like so many other cotton mills around here.

Sunday morning. I jogged halfway up the twenty-one locks to have a look, and then we joined the congregation at Wigan Parish Church. They were in vacancy, with their vicar having just moved on, so a past vicar was there to preside. The congregation was elderly, and friendly—with each other, and with us. The craic was irrepressible—even through the anthem (sung excellently, though the choir was only four). "Irrepressible" is a good word for Wigan. It's a confident town, in spite of the tough conditions. You see it in the way people dress—there's defiance in the air. Whether it's the old-timers seated around us, who were christened and married and had seen out friends and spouses at church, or the younger generation on the streets, as the nightlife starts to gather momentum.

Sunday, after church, and we each took ourselves off for a walk to explore. Jen, off along the canal in the Liverpool direction; me, to find a place to eat that evening. One of the liveliest pubs, full of people of all ages—which is always a good sign—was The Moon Under Water. "You were brave, eating there," said Eth later. The pub's name was chosen by Wetherspoon's chairman, Tim Martin, after he heard that George Orwell wrote a *London Evening Standard* article in 1946 about his imaginary favourite pub: The Moon Under Water.[134]

Nay, lad, that's Wigan Pier tha' cun see

Orwell spent time in Wigan in 1936, while commissioned to write about poverty in northern towns. *The Road to Wigan Pier* (published in 1937) received mixed reviews at the time but has become a classic of social history.[135] I tracked down a 2011 article from the *Guardian*, marking seventy-five years on from Orwell's observations. Wigan Pier is just down a lock and under a bridge. It used to be a heritage centre but is now closed.

The *Guardian* article told the story of why it was called a pier:

> In fact the "pier" never existed, except in song and laughter. The story goes that day-trippers on the train to Southport, peering out across the blighted landscape in a thick fog, spotted a railway gantry leading to a jetty from which coal was tipped into barges on the canal. "Are we there yet?" asked a passenger, mistaking the ghostly outline for one of Britain's newly fashionable seaside attractions. "Nay, lad, that's Wigan Pier tha' cun see," replied the railway signalman. True or not, the pier became a music-hall staple of George Formby.[136]

The council are trying to make Trencherfield Mill a heritage centre now. But why would people want to be reminded of the past? Especially when there's beer and humour to be had.

There are sad places—the Roman Catholic Church of St Joseph, for instance. The door has been broken open, the padlock hanging useless. Inside it makes me want to weep. There was a massive seminary here too, closed in 1992. A statue of St Joseph remains, looking over the place: what must he have seen over the years.

This massive culture chasm

There was a palpable chasm between the folk in church on Sunday morning, and the folk on the streets as Wigan Parish Church rang out the bell for evensong.

It's more than being "relevant" or "accessible"—this massive culture chasm. Does the Parish Church face the same future as the Roman Catholics? The Church of England too? We're an institution that needs to learn to breathe under water. Like the moon, perhaps. The more we focus on decline, the more attention we draw to it. Better simply to get on with serving the world in whatever way seems appropriate in any given place and time. How better than waking up to the deep psychological distress of climate fear and offering a way through, from lament to hope?

Goodbye to Jen

Bank Holiday Monday, and it's just over three weeks on from the party at Prickwillow—Jen's day for heading off to London and Morocco. First, we walk up the Wigan Flight as far as Lock 73 (they're numbered from Leeds). It's her last chance to say goodbye to the lock gates she has strenuously pushed, pulled, and cajoled in so many ways over the last couple of weeks. She does look good on it though. She told me that it had been tough at times. She hadn't quite known what to expect. "But look, I've basically loved it, Frances," she said. A day or so later she emailed me from London:

> Just visited the London Canal Museum. Very interesting despite the fact I know everything that needs to be known about canals!

After saying goodbye, I walk the towpath, thinking over my time together with Jenny; wondering when I'll see her again. It's strange to live so closely with someone, and then they're gone, and so far away.

The surroundings are urban and feel shabby. I'm now a single woman, and I'm on my guard. Gone are the flowers that were in such abundance earlier. There are different flowers out now: the elder is in bloom; dog roses. I lament the report on the news that there will be more plastic bottles in the sea than fish by 2050, as I see a plastic bag in a tree. Jen would have enjoyed the name, "witches' knickers".

Then, back on *The Lark Ascending*, I see a narrowboat, pulling a butty behind, emerging from the lock on its way to Liverpool. Helen emails me from *The Jam Butty* later:

> I wish we had had time to stop, moor up and chat but we have to be in Liverpool for Wednesday so we're on a bit of a timetable. I've found your blog, have read a few posts, and will return to read the rest. Just read about your new haircut and, funnily enough, I noticed how fab your hair is while we were passing!
>
> We are Christians too, C of E to boot. Our home church is Aldridge Parish which is a short walk from our home in Aldridge. We live in our house for the winter months and then summer

> on the boats, making & selling jam to earn a bit of income. When travelling I spend a lot of time worshipping God in the scenery. We use the Boaters' Christian Fellowship website for the comprehensive list of canalside churches and try to get to services when we can. We've been to a wide variety and it's lovely to find God in such diversity.
>
> Our jam business has been going for six years now following a major detour in my career path. We love it. Our website is wildsidepreserves.co.uk.
>
> As well as [making] jam, we both write too. Andy (my husband) earns more than I ever do as he writes for *Waterways World*, usually publishing about six articles a year. He also blogs at captainahabswaterytales.blogspot.com and is way too fond of alliteration for my liking! I write for myself, mainly. I blog at gettingabreastofthesituation.wordpress.com which is primarily a breast cancer blog. I write poetry, mostly bad poetry, but I do love to dabble.
>
> We will be leaving Liverpool docks on Sunday 3rd June and heading along the LL until Leeds so maybe our paths will cross again. If not, I wish you every success and much happiness in your new post and hope you have a smooth move.

I hope we catch up with them again.

The following day, Eth and Harry picked me up for a meal. We went down memory lane too, exploring old haunts of my curacy. We saw the house where two of our sons were born—home deliveries both, and both born on Sunday mornings. "Any excuse to get out of church," said Harry. We visited Tilda's first school, and St Bartholomew's Church, rebuilt after the Victorian barn of a place burned down in 1990. That was a sight in the middle of the night! I reckon I saved the tower by insisting the fire officers concentrate their water on the base to stop the fire funnelling up. The Rector at the time did a great job of rebuilding. It's lovely inside.

Craic and defiance

Back to Hart Common, to their welcoming home, for a brew. Then I'm dropped back to *The Lark Ascending* for my first night alone. There's a beautiful full moon over Wigan Lock 87, and I am stirred to read Psalm 77, with its plea to God from the depths of night. That I'm not alone, that these feelings go way back to the beginning of history also comforts me:

> I cry aloud to God;
> I cry aloud to God and he will hear me.
> In the day of my trouble I have sought the Lord;
> by night my hand is stretched out and does not tire;
> my soul refuses comfort.
> I think upon God and I groan;
> I ponder, and my spirit faints.
> You will not let my eyelids close;
> I am so troubled that I cannot speak.
> I consider the days of old;
> I remember the years long past;
> I commune with my heart in the night;
> my spirit searches for understanding.
>
> *Psalm 77:1–6*

There is so much that is buoyant about the human spirit. That survivor spirit of Wigan, with its craic and defiance, its solidarity of kinship and sense of all being in this together, has made me raw. It's different in the north; somehow closer to the edge of things, more real. God feels very close; the rumour of God still alive. There are folk around who would look out for others, for me, if I needed it. There's a resistance to the worst excesses of neo-liberalism, with its atomizing individualism, turning us all into commodities that seek satisfaction in consumerism. Of course, there's lots of that here too, but I sense something deeper, a knowledge of trouble and the importance of the comfort of strangers and neighbours.

What would happen if a town like Wigan embraced the ideas of Dieter Helm? If the town brought that imagination and survivor spirit to leading the way? Dieter Helm says there is a great deal you can do with brownfield sites in and around major cities to create wildlife corridors—he mentions the London Wetland Centre that opened in 2000, created from four disused Thames Water reservoirs.[137] This area would lend itself brilliantly to his ideas of natural capital regeneration, where he proposes a genuinely green Green Belt for every major town and city, as part of a network of Green Belt National Parks. "The green wildlife corridors would explicitly plan in greener canals and canal paths and green railway corridors."[138] He says there is a host of ways this can be done, and to do so would contribute to a national plan that coordinates current spending on the environment, bringing together a Nature Fund that would ensure pollution is properly costed, which would, in turn, effect a dramatic change in farming practice and resource efficiency. He calls for a radical change of direction to pursue net environmental gain (rather than the current policies which result in net environmental damage), in order that we look after the future.[139] His idea is of a national Nature Fund which would be answerable to government, would manage its own budget, and could also own assets, much as the National Trust, the Royal Society for the Protection of Birds, and the Wildlife Trusts do, owning assets-in-perpetuity.[140] He concludes *Green and Prosperous Land*, his excellent blueprint for rescuing the British countryside:

> When people in 2050 look back on us now, what will they say? Will they say that we met our obligations, were good stewards of the natural environment, and bequeathed them a better set of natural capital assets? Or will they say we promised but did not deliver? . . . It is up to us, and it is not that difficult. It is in our collective economic interests. Yet the opposing forces are powerful and organised, while the conservationists are less so. Facing up to the intensive farmers, the agrichemical companies, the house-building companies, and the transport businesses, the mass membership of nature organisations punches way below its weight . . . .
>
> When people look back in 2050, it is possible that they will see this as the time we stopped going backwards and got firmly on the front environmental and economic foot. They could look back at the 2011 White Paper on "The Natural Choice" as the

> moment when two core ideas took root: that it is the duty of any generation to look after its natural capital so that it is passed on in better shape to the next; and that no economic policy makes sense unless the environment is at the heart of it, rather than as a separate silo of nice things that might be afforded if the growth of the rest of the economy makes us rich enough to care about them.[141]

Ever since reading his book, where he quotes "Big Yellow Taxi" by Joni Mitchell, the words have been rolling around in my mind: "Don't it always seem to go / That you don't know what you've got 'til it's gone? / They paved paradise and put up a parking lot."

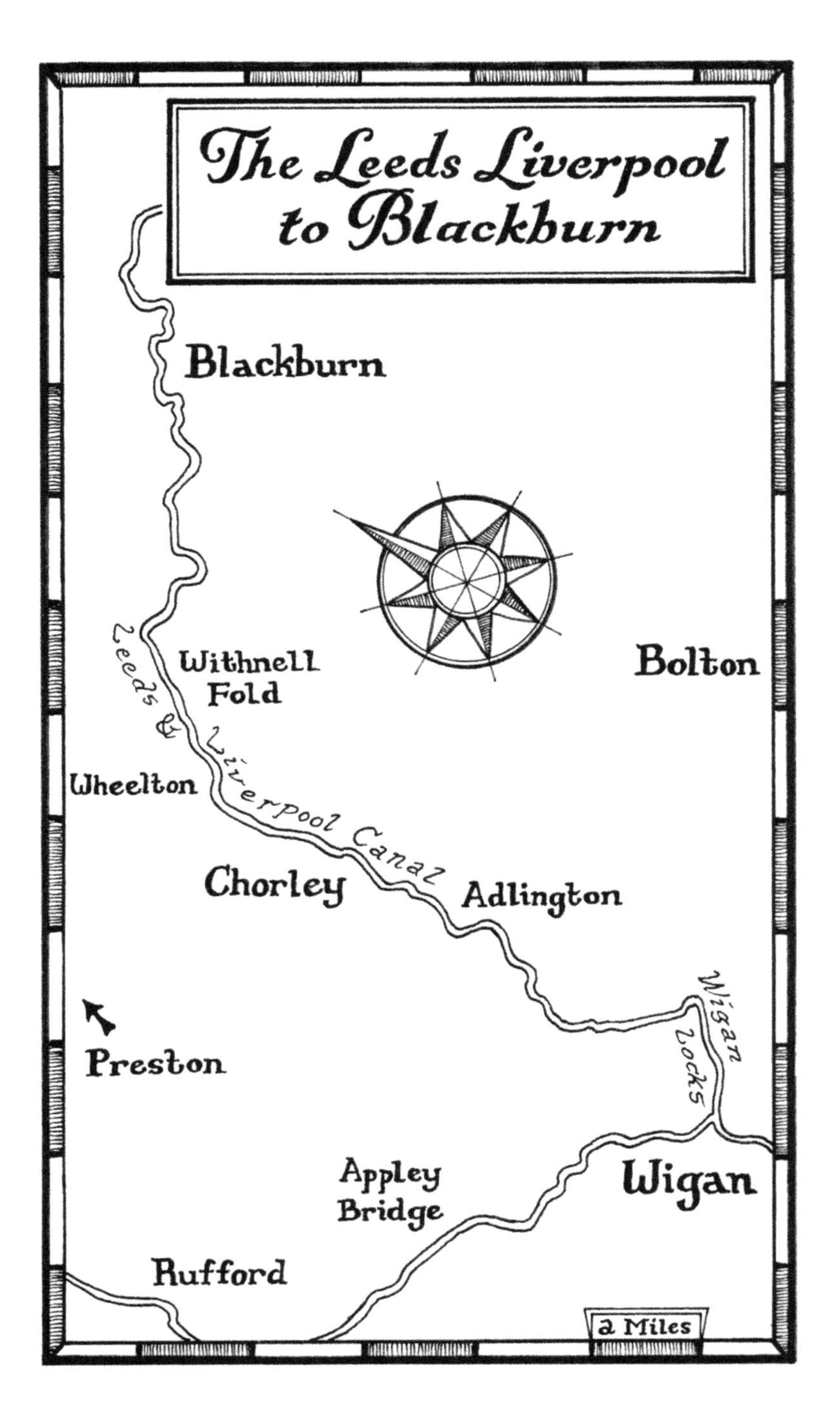
The Leeds Liverpool to Blackburn
Blackburn
Withnell Fold
Bolton
Leeds & Liverpool Canal
Wheelton
Chorley
Adlington
Wigan Locks
Preston
Appley Bridge
Wigan
Rufford
2 Miles

CHAPTER 11

Up the Flight

So now, what with the removals firm going into liquidation, Peter and I are live-aboard. We have no other bed, or hole, or nest to lay our heads. There's nothing for it but to take our time and enjoy the Leeds and Liverpool over the next week or so, while we wait for the British Association of Removers' Advanced Payment Guarantee scheme to kick in, so another removals firm can do the job of bringing our furniture from a warehouse in Thetford to Workington. The date offered now by the latest firm Peter's found is 18th June. We're not anxious.

Off to Mirfield for the end of term

Thursday—the final day of term—was Corpus Christi, and the Mass at noon found us worshipping God in the beautiful church belonging to the Community of the Resurrection, as Father George presided and admitted my friend Mark into the Society of the Resurrection.

Fine singing from the Schola (the choir), and Father Peter playing Messiaen for the voluntary. The Messiaen was nothing I'd heard before. Sombre, intriguing, unresolved. Father Peter and I talked afterwards about how Messiaen knew the monastic traditions of chant that have continued—as they do at Mirfield—since the seventh and eighth centuries. Messiaen included birdsong to capture the natural and unselfconscious theology of premodern music and nature. We talked of that tension between the conscious intention of a composer, since the Enlightenment, and the premodern traditions of monastic chant, in which one simply loses oneself in singing that is done to the glory of God, singing which speaks of the self, lost and found, in music.

The Christogram at the foot of the wonderful icon above the Upper Church altar catches my eye. I'm going to embroider it onto Peter's stole. There was a blessing of stoles at the end of the Mass. Peter's isn't finished yet, so Father Peter blessed some embroidery silks, and a fine piece of gold linen. The blessing will permeate the rest—when it's done.

Peter has made many friends during his short nine months at Mirfield. It has been a rich and wonderful year. The church will be very well served by those he has trained with, prepared in patterns of prayer and the disciplines of holiness.

◆ ◆ ◆

With hired van packed to the nines, and the car as full, we set off on the M62, then the M6 to Penrith, then the A66 via Keswick to Workington. Our new home-to-be. And St Michael's Church. Our daughter Tilda joined us for cake and grapes on the back lawn, and we talked and caught up with news as we pulled dry moss out of dry grass—an unusual drought for West Cumbria—before Peter and I set off for Wigan.

The Lark Ascending is as we left her, on the Saturday morning.

Today we head off towards Liverpool for the day—to Parbold. "Not parboiled, Peter! Don't be silly!" It's lovely having Peter here, getting used to *The Lark Ascending*. The gentle, almost imperceptible rocking and the loveliness of the space holds us with a sense of adventure for what the next days, weeks, months, and years hold.

A more humane future?

Foremost in my mind is just what sort of ministry the Church can offer to those who are overwhelmed by eco-anxiety. How can faith in God, who creates, redeems, and sustains the Universe, enable a more humane future, where God's grace enables us to be entangled with nature, at one with the Spirit who holds all things and beings in a glorious order that reflects God's glory? What can I offer, as I look to the future? How can this experience of working with my own deep anxieties, desires, lament, and fierce hope be offered in the service of the God of faith, hope, and love?

I bring to mind a familiar icon of Jesus Christ, with gentle, challenging eyes, and know his questioning on my life. That gentle challenge needs steel.

It was the fourth of June. "You going up today?" Light Mancunian accent . . .

The locks are large and deep

It was early—dog-walking time—and with Peter driving, I was doing the couple of bottom locks in Wigan. We were about to meet Harry and Eth at the large pound at the bottom of the Wigan Flight. They were joining us for the day, to help with the twenty-one other locks that have the reputation for being the hardest in England.

It's always better to pair up—the locks are large and deep. So Keith was as keen as I was that we should find each other. He and Alison come from Ashton-under-Lyne and keep their boat at Scarisbrick. They had just set off to cruise until October. Semi-retired, Keith told me he'd just turned down the first job since 2002: "Painting a bloke's bathroom ceiling. Just the sort of job I like. Brings in the beer money."

Harry and Eth joined us. They'd met the volunteer from the Canal and River Trust as they walked down from where they'd parked the car at Rose Bridge, halfway up. He'd told them there wasn't enough water in the pounds—particularly between Lock 80 and 81. So we'll have to wait until it fills up, when we get there.

The first four locks—85 to 82—went without a hitch: slowly and carefully, as Harry and Eth learned the ropes from Keith and Alison. Keith and I got to know each other as we chatted about life. He'd had bowel cancer. Life meant more to him now. He didn't know quite what to make of Peter's and my life changes, so we talked of his boat, and of the Marple Flight on the Macclesfield Canal—one of his favourites—and of where they planned to go. Leeds. York. Oxford. Wherever.

When we got to the pound below the empty one, we moored up to have a look. The mud, rocks, and weed, and other things that shouldn't be in a canal, were there for all to see.

Water? Not enough

The CRT volunteer told us not to go if the level was more than a foot below the overflow. He arrived, and said he reckoned there might be enough. Our risk, of course. When we did go, "Best to stick to the middle where there's most depth. And if you stick, just wait until there's enough to float off." Obvious, really. As if we'd have a choice.

Keith went first. All well until about halfway, then he ran aground. Following behind, I slowed *The Lark Ascending* right down, and so lost steerage, ending up diagonal, with stern and bow in mud. And a nasty sound under the propeller that made me think of shopping trolleys. Keith pushed on a little further, slewing off to the right. And stuck again—a rock this time? There wasn't much more water at all, but enough for me to come alongside and pass him, picking up his bow line as we went, which pulled him off whatever it was, and into the lock we went. As the lock began to fill, "That's enough drama for the day," we agreed. This was number 80.

People told us there were three boats coming down. A singleton and a pair. Good news, as they bring water with them. The singleton came out as we went into Lock 78. Then that sound below that you quickly learn to dread—and don't expect to hear in a lock, where there's usually enough water. I put the engine into gear, forward. It stalled. I started it, and into reverse. It stalled. Keith said he could see something white below the rudder. "Best not to try and clear it with the engine. You'll need to go into the weed box." I said I hadn't done that before.

The pair were waiting in the pound—the rest pound between Locks 78 and 77—so Keith towed me into the next lock. The silence was wonderful, without an engine. I began to imagine the world of horses; the slipping through the water; the noise of iron shoe on stone cobble. A horse can pull a ton on the road; it can pull a hundred tons through water.

It wouldn't have been silent here, though. Just different sound. The noise of the collieries all around. The Nicholson guidebook tells of this flight as an industrial hub with collieries and ironworks lining the canal.

A way of life—a way of death

The Rose Bridge Colliery—just where my engine stalled—and Ince Hall Coal and Cannel Company would have been pumping out noise and sulphur all around us. (Cannel was a dull coal that burned with a smoky, luminous flame—so the guidebook says.) The Wigan Coal and Iron Company was the biggest, employing 10,000 people at their works all alongside the top nine locks of the flight. It owned pits all around this area.

It was one of the largest ironworks in the country, mining a million tons of coal to produce hundreds of thousands of tons of iron a year, with the skyline dominated by blast furnaces, coking ovens and chimneys. There wouldn't have been much silence, day or night, with the Industrial Revolution establishing our reliance on fossil fuels as a way of life—a way of death. As Dieter Helm says, "Prosperity came, built on a fossil fuel economy, bringing with it pesticides, plastics and petrochemicals."[142] Air that killed.

Chances are that a hundred years ago, the collieries and works would all be just starting up again, after the Whit Weeks. They took it in turns to close—in Bolton, Farnworth, Leigh, Wigan—to give the workers two weeks off to cycle or walk to Blackpool for the annual holiday. When I was curate in Westhoughton—only a few miles away—it was still a fresh memory for the older folk. The Whit Walks still happened then, though with the closing of the last collieries, in the eighties, the ceremonies—the dressing-up of rose queens, and her bridesmaids carrying baskets, dressed all in white—were not going to last much longer.

◆ ◆ ◆

We paused in Lock 77. With Harry, Peter, and Keith looking on, I undid the nuts (with my old cycle dog bone spanner) that held the weed-hatch in place. The propeller was shrouded in white and black—some fabric or other. I made sure the key was out of the ignition, and gingerly felt down. Tight-wound it was—and there was electrical wire there too. Now armed with scissors, I started to cut wherever I could, and gradually it came free: a sturdy black-and-white striped shower curtain, still with rings.

The whole lot came up. Two lengths of cable and a long gauze bandage, which must have been there for a while.

"You need a better spanner than that," said Harry as I tightened the box lid down. "I'll bring one when we meet in Skipton." Keith suggested putting the engine on and into gear, to make sure there was no water leakage. All good, and good to go.

I wish I'd taken a picture of that shower curtain from Wigan. I did later, though, of other stuff that had been fished out of the canal. Rolled-up carpets. Bikes. Baby walkers and the inevitable shopping trolleys. Such a lot of rubbish. We are such pollutors; so uncaring of the beauty of the world.

Dieter Helm says that making polluters pay is the single most radical and effective policy that could be adopted, both for economic prosperity and for the environment:

> The British countryside would be radically different, and radically less polluted, were this simple economic principle adopted. It would not cost anything to the economy in aggregate, and at the same time it would yield lots of revenue, some of which could go to repairing past damage and enhancing our natural environment.[143]

Once the other boats had passed us downward, and with no more coming down, and with more water in the system gushing down the bypass culverts, or by-flows (as the CRT volunteer called the side channels that supply the holding pounds, with the flow controlled by sluices upstream), Eth and Alison were in their stride, opening lock gates way ahead of us. Leaving Harry and Peter to close up behind us, Keith and I took each lock in turn, relaxing into banter as we went.

He said he'd sell up the house tomorrow, but Alison liked her bricks and mortar. "Though I'm surprised how well she's taken to it." Forty years married, they hadn't any children, but had always had a dog. He was worried that Dolly—eight months old, and a cavapoo—was showing signs of attachment anxiety. She'd had a previous owner, who'd paid a mint for her and had kept her in a kitchen cupboard, under the sink. Dogs aren't accessories, we agreed. (Though Dolly looked like one—pretty, and silly.) Alison caught her up into her arms when Diesel came along. Diesel, explained his owner, was an American Staffy. "Most owners around here don't treat them well. Make them aggressive. He's as soft as anything," he said. "There's dogs like him that would have that dog for breakfast. But he won't. He's as soft as anything," he repeated. Again and again, he reassured us Diesel wasn't a pit bull. He was, though.

"You're not far from the Fells, though"

Once we were in our rhythm, Keith wanted to know why Peter was on the phone. I said we were liveaboard now, as it was taking time to sort out our move to Workington before he was ordained. We talked of life changes. That Peter had been a children's doctor all around here, based in Bolton, for years. That I was a priest; that I knew Manchester well. That we were both looking forward to Workington. Steve looked doubtful. He nodded, still unconvinced, when I tried to explain how Peter and I felt we wanted to be where we could make a difference. He nodded, again with not much comprehension, as I mentioned coastal town poverty and the West Cumbrian coast. "You're not far from the Fells, though," he reassured himself.

We were in the lock just below the top one (which had recently been renovated). It looked beautiful, compared with the dilapidated state of some we'd been through. If only there were money for the rest. "It used to take three days to replace a lock gate," said Harry, who used to work for the Environment Agency. "When there were lots of lock makers, and British Waterways were in charge. Now it takes six weeks if you're lucky." Keith grumbled it was now the CRT, "but they've just spent too much money on rebranding."

Just before Top Lock was the Kirkless Hall Inn. "I'm back there for a pint," said Steve. They issue certificates. So, of course, we joined him. I made haddock fish pie followed by strawberries and cream for lunch, once we'd moored up. I felt the decades fall away again as I went to prise Peter, Harry, and Eth out of the pub, once it was ready. A fishwife, me.

◆ ◆ ◆

After lunch, we walked back with them to the car at Rose Bridge, marvelling that though we hadn't seen each other for years, the friendship was as fresh as ever. Walking down that flight, now full of water, took no time at all, compared with the five hours it had taken to travel up the three hundred feet or so. Once Peter and I were back on *The Lark Ascending*, we headed onwards for a couple of miles to Haigh Hall: somewhere we used to take the kids for a day out in the early 1990s. It's a stunning location, with views extending across the Cheshire Plain to Runcorn, and beyond to the hills of Wales. Wigan was below us now, surrounded by green woods and fields.

"Haigh"—the word comes from an Old English word meaning "enclosure". This is long before the enclosures of the eighteenth and nineteenth centuries. There would have been a timber-framed manor house here in the late twelfth century, when the Norreys, originally from Normandy, held the manors of Haigh and Blackrod. Hugh le Norreys had a daughter, Mabel, who became heiress to the fortune. We moored near Lady Mabel's Wood. Sir William Bradshaigh married her in 1295. The Bradshaighs held the estate until 1770, when it passed to a niece, ten-year-old Elizabeth Bradshaigh Dalrymple, though held in trust until she married her cousin Alexander Lindsay, 23rd Earl of Crawford, 6th Earl of Balcarres. Alexander sold the Balcarres estate to his younger brother to fund repairs to the hall, which had not been lived in for many years and was damaged by mining subsidence.

Volatile cotton markets

The house was rebuilt by James Lindsay, 7th Earl of Balcarres, creating the building of today, with extensive parklands and gardens and a purpose-built curling pond which rivalled Scotsman's Flash. Sandstone was bought from Parbold on the Leeds and Liverpool and dressed on-site. During the 1860s, forty miles of pathways were built by local Wigan men, many of whom would otherwise have been destitute, plummeted into poverty by volatile cotton markets. I thought of them as I jogged those same pathways the following morning.

The Lancashire Cotton Famine of the early 1860s followed the boom years of 1859 and 1860, leaving families in extreme poverty. Overproduction flooded the market with finished goods, while raw cotton was in abundance. Demand fell and prices collapsed, with the market further complicated by the interruption of baled cotton imports during the American Civil War. To their credit, many Lancashire cotton workers—despite the real hardship they suffered—resolved to support the Union in its fight against slavery. In the name of the working people of Manchester, they wrote to President Abraham Lincoln of their hope that "the erasure of that foul blot on civilisation and Christianity—chattel slavery" would be seen during his presidency. He responded within a few days in January 1863:

> I cannot but regard your decisive utterances upon the question as an instance of sublime Christian heroism which has not been surpassed in any age or in any country. It is indeed an energetic and re-inspiring assurance of the inherent truth and of the ultimate and universal triumph of justice, humanity and freedom.

There's a monument in Lincoln Square, Brazennose Street, Manchester, to commemorate this international friendship, upon which parts of both letters are inscribed.[144] All around here, throughout Lancashire, workers became unemployed, and went from being the most prosperous workers in Britain to the most impoverished. Many emigrated. My grandmother's maternal family left from Manchester for Australia in the 1860s.

Haigh Hall once housed a magnificent library, the "Bibliotheca Lindesiana", which was gifted to the Rylands Library in Manchester. The house was opened as an auxiliary hospital for convalescing soldiers in November 1914. It was sold to Wigan Corporation in 1947, when the Lindsay family returned to their family seat in Fife, Scotland. Now it's used for weddings and functions. "It's a sad place," said Peter, as we wandered around—the house, perhaps, but the woodland around is one of the largest and most ecologically important in Greater Manchester. We saw a toad—the first either of us had seen for years—as we walked down to Sennicar Bridge, Canal Bridge 61, and home for the night.

Trees. Woods. Forests.

It was as Mary Colwell describes it:

> There is something tinglingly magical about woodlands, even small patches . . . They are steadfast and full of expectation; there is a sense of a change of shift from day to night, from the known, visible world to the realm of covert creatures that move in shadows. After the wet, warm winter the rotting leaves and rich soil give off a primeval, earthy smell. As a cold wind buffets the valley, the trees provide a sense of calm. I feel I am wrapped in a woody blanket.[145]

Oliver Rackham is optimistic about woodland. He reckons that it's reviving, due not least to the success of the Woodland Trust, founded in Devon in 1972, and a change of direction in the Forestry Commission, now fully committed to woodland conservation.[146] Woods and forests are now valued, and the ability of trees to capture carbon is accepted widely. Cost? It's estimated that a trillion seedlings, which would remove two-thirds of all the carbon dioxide that human beings have ever released into the atmosphere, would cost a minimum of £240 billion. As Andrew Davison points out in a 2019 *Church Times* article, that's the cost of two Apollo space programmes. He comments that "to put the finances into perspective, the two richest people in the 20th century could each have paid for it on their own . . . The combined wealth of the billionaires of [the US] alone could run to it 30 times over" and writing in celebration of the Moon landing fifty years on, he concludes, "To attempt the Apollo programme was an act of magnificence. An attempt to reforest the planet would be similarly magnificent; indeed, not to attempt it would be madness."[147]

But first we must stop the burning of the rainforests, particularly—but not only—in the Amazon basin. Not only are we witnessing the destruction of countless species of plants and animals (not to mention the peoples who live there in symbiosis with their environment), but the rainforest also absorbs millions of tons of carbon. When the trees are burned, the carbon and carbon monoxide released have a global impact.

The moisture and water that the trees absorb mean these enormous areas are like reservoirs, cooling the earth and keeping it fertile. The result of destruction, where rainforest is cleared, will simply mean desertification, not fertile arable land.

Trees. Woods. Forests.

I wander off that evening, unable to cope with the despair I feel. It burns within me, making my muscles tense with frustration and helplessness. I want to scream. I can even see why people self-harm, though that's never the answer. The pain is intense as I imagine the forest all aflame, creatures destroyed by fume and fire, indigenous peoples deciding not to run. Where would they go? I don't know how to contain this intensity within me:

> I am utterly bowed down and brought very low;
> I go about mourning all the day long.
> My loins are filled with searing pain;
> there is no health in my flesh.
> I am feeble and utterly crushed;
> I roar aloud because of the disquiet of my heart.
> O Lord, you know all my desires
> and my sighing is not hidden from you.
> My heart is pounding, my strength has failed me;
> the light of my eyes is gone from me.
>
> *Psalm 38:6–10*

As it gets dark I lie on my back in a clearing in the woods and watch the high clouds through thin branches. It is so quiet. I listen, and hear the silent sobbing of the earth.

CHAPTER 12

Botany Bay

That sense of lament continued with me. We were in the heart of Lancashire, where the history was hard and resonated with my heaviness.

I'd driven up the M61 many times. You know you're passing Chorley by two distinctive landmarks. One is the tall spire and imposing Church of the Latter-day Saints to your left, the Preston England Temple. It was dedicated in 1998, the fifty-second such temple across the world, an impressive complex of Olympia white granite from Sardinia with a zinc roof. The site, which includes a missionary training centre and a family history facility, is the largest Mormon temple in Europe, serving Latter-day Saints from the Midlands and northern parts of England, the whole of Scotland, the Isle of Man, Northern Ireland, and the Republic of Ireland. It's a symbol of thrusting religion, planted with little sensitivity to its surroundings.

The other landmark is a converted and restored canalside cotton mill, now a shopping and entertainment complex called Botany Bay. The original mill was built in 1855 for Richard Smethurst, son of an early pioneer in the Chorley cotton industry. Despite the Cotton Famine and temporary closure in 1861, Canal Mill continued to manufacture until the end of the 1950s, when it eventually closed for good. In 1994 a local entrepreneur purchased the mill, and after a complete renovation and restoration, "Botany Bay" opened on 1 December 1995. It features five floors of shopping, a garden centre, restaurants and coffee bars, and an indoor play centre, drawing people from near and far. Peter's idea of hell. "Why 'Botany Bay'?" he asked.

There were several mills here from the late eighteenth century, as the area developed to become the main port for Chorley. The nearby Lancaster Canal was begun, originally intended to run from Walton

Summit to the Bridgewater, but it never happened, following the building of a temporary, and then permanent, tramroad to connect Preston to the rest of the system. Botany Bay was where the Irish navigators (or navvies) lived. The canals they built—sometimes known as "navigations", or "eternal navigations"—were intended to last forever.

By the 1830s most navvies were building railways

It's now recognized that the great majority of navvies in Britain were English, with only 30 per cent Irish, but the prejudice stuck. Locals saw the Chorley site as an area to be avoided, much as you'd avoid the penal colony in Australia. Hence the name. I imagined Joseph Banks seeing a cove on the Australian coast and naming it after the flowers he loved so much, and all the new varieties that delighted him. So innocent, belying the truth of the brutality, labour, ugliness of those early years of that convict settlement in New South Wales. When we hear mention now of Botany Bay, flowers are the last thing to come to mind.

Botany Bay wharf, here in Lancashire, became an important hub for traded cotton, transport, and communication, with services running to Manchester, Wigan, and Liverpool. When the Lancashire Union Railway opened in 1869, it ran through Botany Bay and over a viaduct across the canal, and began to supersede the canal in coal transport between Wigan and Blackburn. The railway, in turn, remained in service until the 1960s—the viaduct was demolished for the construction of the M61 in 1968. Canal, rail, road. The transport system changed fundamentally as economic forces became global, reaching into ever-expanding markets and trade around the world.

Godforsaken

The history of disruption, of communities on the move, of brutality endured by the navvies who worked, far from home, in grim conditions, constructing the canal we were now enjoying, deepened the dark place I occupied internally. I felt bleak and dispirited, unable to shake off the sadness, frustration, anger pulling me down. I felt godforsaken; that the world was godforsaken. It occurred to me that my sense of foreboding about climate change was not going to go away. I wasn't ever going to recapture the sense of innocence and peace of mind that had been mine, as I remembered it, as a child. That same innocence would be claimed from each young person in the future—we would all need to learn to live with this state of constant swing between lament and fierce hope that was a new normal. The story of Greta Thunberg, of her plummet into depression when she was eleven, her inability to speak, to go to school, until she decided to go on strike in August 2018, is the new normal. She describes the transition from depression to activism. It's the journey I'm on too.

I went for a walk, back along the canal. I needed to be alone. Perhaps it was because we were now in familiar territory, where we had lived before and begun family life, with small children to bring up—where Peter and I had both worked at a significant time. Perhaps it was because it was just us now, as we began a new chapter in our lives. It seemed, suddenly, that we had left the south behind; that the north was a reality. The past was behind, the future ahead. This felt like a turning point on the voyage.

I found myself wondering whether Jesus knew what the future held as he cried out from the cross, "My God, my God, why have you forsaken me?" He was in the utter depths at that point—his life, healing, teaching, ministry, all seemingly worth nothing as he died. And who knows now what lies ahead for planet Earth? For the humanity that has done such damage? A sense of foreboding was deep within my soul. To remain there, though, is to remain on the cross. With all its reality, it is not the final, tortuous resting place of Christ. He descended to hell, yes; but he didn't remain there either.

The narrative of Jesus Christ continued and continues today. Mary Magdalene meets her gardener at dawn on that first day, surrounded by the green-gold light of resurrection, the birdsong, the flowers. Then there were conversations in upper rooms, on the road to a village, and fish eaten by the lakeside. This was new life that showed a love stronger than death, hope fiercer than the tomb. There was an ascension into the clouds of mystery and unknowing, but along with that going, a leaving behind of the promise of the Holy Spirit, until the end of time. That

Holy Spirit is here yesterday, today, and tomorrow, inspiring with love, breathing through all creation the life force of God. The Holy Spirit, the great encourager, gives heart and hope to everything that exists and is the positive energy that transcends death and depression, bringing hope. As I realized that for the rest of my life I would plummet into depression and despair, so also I believed I would be inspired by hope. The lament, too, was necessary—the expression of the pain of grief and intense fear for future generations, in what seem increasingly likely scenarios. What I must do is accept this new and complex realization, and continue to live as if there's no tomorrow, with a fierce hope and a fullness of engagement and action that transform the pain into something positive. The Cross is the tree of life.

◆ ◆ ◆

When I got back from my walk, we were given a lift into Chorley by folk who lived aboard a widebeam moored just under the motorway. They were off to the market for Rufford new potatoes. "Just rub them and the skin comes off a treat," said Jed. Enjoying those new potatoes delighted us that evening.

"I couldn't stand the noise," said Peter, as we had a pint of Wainwright on the way home at a pub just under the M61. We drank from Wainwright glasses that instructed us to "find your mountain". Alfred's words were there—the latest manifestation of the marketing phenomenon he has become—ironic, given his churlish misanthropy. "You were made to soar, to crash to earth, then to rise and soar again," the glass told us. This we *must* have. "I didn't pay for them; why should you?" responded the publican, as she gave me two clean ones to take away. We surmised that wouldn't happen in the south. Wainwright came from Blackburn, where we'll be in a day or so. He found his mountains in Cumbria, falling in love with the fells at first sight. He took his Lancashire soul with him, though, when he moved to Kendal.

Lancashire fracking

Not far from us, at Preston New Road, there are plans to start fracking—a process where water and chemicals are injected into rocks at high pressure to extract gas. Oh, how tired, frustrated, angry it makes me! Human rapaciousness: will it never end? Burning rainforests; blasting rocks. It makes so little sense to develop fossil fuel extraction and technology when we are experiencing the hottest weather ever because of the carbon gases that are surrounding the planet, trapping so much heat that wildfires rage and harvests fail. Chaotic weather already threatens the production of grains and open-air vegetables.

Fracking takes us in entirely the wrong direction. We should be capturing carbon from the atmosphere—and the best way to do that is by planting trees. Where's the investment into renewables, wind and solar farms that offer the added bonus of habitat for wildlife, and reflect the sun's rays as they generate energy? We need to rely less and less on fossil fuels, and as quickly as possible, particularly as the fracking process also releases methane, so much more dangerous than carbon dioxide. Fracking risks poisoning the water table as well. It makes no sense.

Peter and I used to live nearby. This part of Lancashire has its own spirit. The market was on in Chorley—the "Flat Iron" Market which dates from 1498 (named that either because the weavers used to hold down their wares with flat irons, or because the space it used to occupy was shaped like a flat iron—or both). We wandered through, hoping for some remnant of the textiles and haberdashery trade that would provide me with some further material for the stole I'm making for Peter's ordination. No joy. Hopefully Blackburn will provide. Time's running short, with the ordination on 30 June.

A good bookshop, though—one to browse in. Something caught my eye.

One of the reasons—I reckon—that I love being on water is that I went by ship three times from Australia to England when I was little. I was seven on the third trip in 1966. The three liners we sailed were the *Castel Felice*, SS *Northern Star*, and SS *Himalaya*.

Many thousands to Australia

And here was a book all about the days of the ten-pound assisted passage that took many thousands to Australia in the 1950s and 60s, seeking a new life far away from the austerities and smog of post-war Britain. The voyage took a month.

The 27,955-ton *Himalaya* was built in Barrow by Vickers-Armstrongs in the late 1940s, was owned by P&O, and from 1958 until 1974 transported migrants and cruised the seas. It must have been in the *Himalaya* we sailed in 1966 on our final voyage to England—for good—because it had a swimming pool. I had my birthday on that trip, and can remember that while Mum and Dad were doing their own thing, I'd have the run of the ship, wandering all over. I watched the swimmers in the small pool, shaped rather like the locks we've been in, with no shallow end and a swell like the sea all around, and thought to myself, "People can swim, so there's no reason why I can't. If I stick near the side, so I can hold onto the rail if need be, and kick and doggy-paddle, I'll make it from one end to the other." Ten yards, maybe. It can't have been far, but it was certainly out of my depth. I taught myself to swim. No one watching. No one to tell me not to.

What would I say to my seven-year-old self? Peter and I play that game in the evening. Back in the sixties, before any thought of climate change. We wonder about the wisdom of telling children too early about climate catastrophe. How much better to have seven years or so of relative security—reading children's books that offer a safe and emotionally secure world. Eco-anxiety will hit soon enough—Greta was eleven—and children will be much more resilient, able to live with hope, if their first years are relatively unburdened. Oh, I fear for children and young people, and what they will live with. How risk-averse parents can be about freedom, the freedom I had to teach myself to swim. How anxious childhood is. How much the natural world has to offer to those who struggle with mental health issues. There's a great initiative called "Green Health", with lots of ideas for church spaces, that might well work in Workington.[148]

◆ ◆ ◆

When Jen was with me, we talked a bit about immigration today, and Australia's draconian reputation. It wasn't always the case. Hundreds of thousands of displaced Europeans and over a million "ten pound Poms" emigrated after World War Two—when the fear of Japanese invasion stirred the Australian government into a policy of "populate or perish". There was a "white Australia" policy in place then, until it was repealed in 1966 and Australia began to develop the multinational population it has today. Then all you needed to be was white, of sound health, and under forty-five years of age.

In the mid-1960s when our family sailed from Australia to England and back again, we were accompanied by "ten pound Poms" off for a new life. It was exciting. I remember the flirtatious charm of the Italian waiters on the *Castel Felice* who served me kippers, and the taste still takes me back; the smell of tinned orange juice as well—even now it conjures up the other smells of the ship. I was dressed up as the Queen of Hearts (I much preferred swimming) during elaborate ceremonies as we crossed the Equator, with an intriguing visit from Neptune. We were all given certificates; I kept mine for years.

We stopped at Colombo, at Mumbai, at Aden, went through the Suez Canal, and I remember the beach at Nice where we swam with nothing on.

I loved being on board ship: watching the flying fish, the porpoises that accompanied us; visiting those exotic places, like Aden, where we couldn't go ashore so the traders came to us, raising their goods from small boats far below. I loved the perpetual motion, the roll at night, the constant power and throb of the engine. All around us the sea, extending to the sky. And now that sea is clogged with plastics—in fifty short years.

Contrails

That was the age when the world opened up to ordinary people. Travel became global, as most families had someone living elsewhere. Very soon, commercial air flights would take off, literally, and holidays in warm places became the norm, with the criss-cross of white lines another way to trap the warmth. The contrails left by aircraft reduce the range of temperatures by contributing to cloud cover during the day, reflecting solar energy. At night, they trap warmth. Overall, the contrails increase the temperature of the earth. We should all fly much less—only if necessary. One thing to do is to take holidays in the UK.

When we arrived in Britain, to counter her homesickness Mum threw herself into the English countryside. We'd go for blackberries, make wine from elderflower, and gorse, with its coconut scent; we'd learn the English birds, trees, and flowers. We'd find out-of-the-way spots for picnics and get lost on walks. She grew to love England almost as much as her native Australia. She read aloud to us, as children: the stories and poetry of her homeland. I remember her reciting Dorothea Mackellar's "My Country".[149]

Australia becomes browner and browner. It's predicted that water security problems are going to intensify. Loss of biodiversity is projected, including that of the Great Barrier Reef and the Wet Tropics. Sea levels will rise, with more severe storms and flooding; fire and drought will cause declines in agricultural and forestry production. All stuff that's happening already.

My childhood seems light years away. Today's children face a future that can only provoke anxiety, such that normal emotional development and all aspects of life are under a cloud that intensifies everything by the threat of catastrophe. It raises acute questions of how to protect children against intense anxiety, whilst being realistic and not in denial about what lies ahead.

Reading children's books

Perhaps, as some suggest, the answer is to read children's books for a moral and emotional knowledge that humanizes us. There are some wonderful children's classics—ones I remember my mother reading to me. Like Rumer Godden's semi-autobiographical memories of a summer holiday in France in 1924 with her mother, captured in *Greengage Summer*, which she wrote in 1958.[150] She remembers the scars of World War One—the craters, the shell-damaged buildings—and finding letters, helmets, and other soldiers' belongings in the trenches. Her mother becomes ill, so the family stop at a hotel near Rouen: a halcyon time for the children, with the countryside golden with summer harvests, and greengages ripening in the orchards. A time of emotional transition for the girls, when they move from the innocence of childhood to begin to appreciate the awareness and knowledge that underpins adulthood. Emotional complexity, beautifully captured.

They run wild, in the heat, and are introduced to passions and emotions beyond their experience, growing through innocence. Innocence and experience—it's a strange relationship between the two. So many today assume that childhood is the state of innocence, left behind as we grow towards adulthood. This view is a legacy of the Romantic Era. It's much more complex. Children are so often experienced beyond what is assumed of them; they understand far more than we credit. Their innocence is held alongside a deep understanding of good and evil, light and dark. And as we grow into adulthood, or contemplate old age, our innocence does not leave us. We can continue to experience the moments of insight we had as children, to be apprehended by the God who delights us in the natural world around. Actively cultivating that awareness of God in the world around, even a mystical sense of God's presence, helps us develop thankfulness—so important against despair. That new, complex normal of an emotional state caught between despair and hope, lamenting fiercely to survive, will need much more work. With the story of Jesus, the Church has a pattern to enable us to move on through the darkness, and it can teach us to lament. The resources are all there.

Off by heart

Learning poetry, psalms even, off by heart is immeasurably valuable: words that shape and frame the imagination for the rest of life. For who can read Psalm 104 and not begin to see God in the natural world, in the patterns of the seasons, in the wildlife and natural common goods of light, water, sound, silence, colour? Or any of the other psalms, framing our emotional response in thoughtful and profound ways? There are rich sources and resources here, to enable lament, emotional and moral resilience to cope with the challenges of life; food for our imaginations that is richer and fuller than the thin gruel of fantasy which is too often what children receive today through their screens.

◆ ◆ ◆

We left Botany Bay behind as we moved on, deeper into Lancashire, a short distance up seven delightful locks at Johnson's Hillock.

The bottom lock goes right as what remains of the Lancaster Canal goes left. If one wanted to get onto the Lancaster today, you would need to continue beyond Parbold on the Leeds and Liverpool, and turn up at Rufford for the Ribble Link. The link is only open fifty days a year, so the Lancaster is even quieter than the Leeds and Liverpool. One day we'll explore.

The countryside falls steeply away to the left now, as we begin to climb the Pennines. We moored at the Top Lock, where Canal and River Trust folk were supervising Prince's Trust volunteers repainting the woodwork and metal of the lock with distinctive black and white.

There's some good information on this flight about the way the horses worked. One bridge has no towpath, so a hook in the wall enabled the horse to pull in the opposite direction, propelling the barge out of the lock. The wear from the rope was there to see. When the horse was towing out of locks going uphill, the tow line was passed around the shaft of the upper ground paddle. The grooves made by ropes in cast iron are beautiful.

A long afternoon in the sun, and then a walk with Peter through Wheelton Village and on in wonderful countryside just under the M61, where a sign for Botany Bay meant we hadn't travelled as far as it had felt. The stream below us has nothing in it—no fish, no dippers—which worries us. I longed for the sight of a dipper. A bird that has always delighted me. A poem, "Element":

> I saw the dipper again today –
> the bright and bonny bird –
> it watched me with a canny gaze
> then curtsied, turned, demurred.

Its element allowed me in –
that joyful little stream –
it stirred, aroused a salty passion –
as vivid as a dream.

The water pounded on my soul
until the bird rejoiced –
it dived within the deep, dark pool
and I was found and lost.

No curlews though

A few lapwings in the field, chasing off the crows—so obviously with young. No curlews though. Mary Colwell has written the most beautiful book about curlews, *Curlew Moon*. The bird's Latin name is *Numenius arquata*, and Colwell explains how both parts refer to "its most conspicuous, long curved bill":

> *Numenius* is the Latinised version of two Greek words, *neos* for new and *mene* for moon, that thin shaving of light that is full of potential. *Arquata* is Latin for the archer's bow; taut and stretched into a smooth arc. *Numenius arquata*, then, is the new-moon, bow-beaked bird.[151]

She tells how the last thirty years have been catastrophic for Northern Irish curlews, and many other farmland birds. Back in 1986, a survey found 5,000 pairs of breeding curlews throughout the province. By 2015, their numbers had crashed to fewer than 500 pairs, and probably closer to 250, a decline of over 90 per cent.[152] Why? Silage. Silage cutting has eradicated ground-nesting birds. Farming machinery destroys eggs and chicks indiscriminately:

> Faced with danger, curlew chicks will sink down into the grass and freeze, where they are killed instantly by the rotating blades. Nests full of eggs are flattened. Those birds that do manage to

> survive the machines will often fall prey to foxes and crows, two species that do well in intensive farmland.[153]

I can't look on a freshly cut silage field any more without thinking of the slaughter.

There are swallows and house martins chasing insects on the pounds between locks, and there's an abundance of flowers. Ragged Robin (not to be confused with red campion); grasses galore, cuckoo flower, meadow vetch. The may blossom is over; now the elder is in bloom. I used to gather elderflowers when Mum made wine.

◆ ◆ ◆

Blackburn is just around the corner. One of these days, Peter and I will do all the cathedrals we can by canal or navigable river.

We pulled up at Withnell Fold, intrigued by what the Nicholson guide had said. It was indeed a village built (around an older farm) in 1843 to support the paper mill founded by the side of the canal by Thomas Blinkhorn Parke. Peter wasn't sure why we were bothering. But as with so many aspects of this voyage, our days and places are full of surprises. For a start, the old aqueduct (now underground) built to bring water from Thirlmere Reservoir to Manchester ran through the village.

The longest gravity-fed aqueduct

It's an almost ninety-six-mile-long water supply system built by the Manchester Corporation Waterworks between 1890 and 1925, to carry approximately 55,000,000 imperial gallons per day of water from Thirlmere Reservoir to Manchester. The first water to arrive in Manchester from the Lake District was marked with an official ceremony on 13 October 1894. It is the longest gravity-fed aqueduct in the country, with no pumps along its route. The water flows at a speed of four miles per hour and takes just over a day to reach the city. The level of the aqueduct drops by approximately twenty inches per mile of its length.

> Water. Dieter Helm describes water as a common good which, in economic terms, cannot or should not be privatized. Water, like air, is rivalrous (if one person uses it, another cannot), but non-excludable (people cannot be excluded from using it according to whether or not they can pay). In his twenty-five-year Environment Plan, water is a key element that contributes to the public good of a better natural environment for future generations: the components of the plan are "clean air, water, beaches and good marine habitats, flood defences, carbon-rich soils and biodiversity". [154]

The horseshoe of cottages that Blinkhorn Parke had built in his model village were constructed from stone extracted from the ground when he dug the lodges (water reservoirs) needed to supply the paper mill.

Blow the wind southerly

The contralto singer Kathleen Ferrier (1912–53) had lived briefly at number nine, after her marriage to Bert Wilson. She first came to public notice when they lived at Silloth, on the Cumbrian coast (where Bert was the bank manager), and she won the Gold Cup at the 1938 Workington Musical Festival. Radio work followed, and her national and international career took off, only to be curtailed by her early death from breast cancer. We caught up with her again, later that day, in the Museum in Blackburn.

It hadn't been a happy marriage. Unconsummated, in fact. Though they kept up appearances until after they were amicably divorced. Kathleen told a friend later that she'd once said to Bert that she'd hoped he'd make more fuss of her. He'd replied, "Why chase a bus when you've already caught it?"[155]

Withnell Fold Methodist Chapel was the centre of life in this little village of about 300 residents. We were shown around by a member of the congregation, who explained that Blinkhorn Parke was Methodist, so there was a reading room in the village but no pub. The reading room is now a private home, so the chapel is used every day for an after-school club and for various functions, events, and organizations like the Women's Institute and Alcoholics Anonymous.

Kathleen Ferrier sang and played the piano there often. We heard more about her later that week, when we had supper with friends, Jonathan and Emma. Emma told us how her grandmother, Beatrice Livesey, a concert pianist, accompanied Kathleen at King George's Hall in Blackburn. Those were days of far more musicality within our national life.

The paper mill stopped work in 1967. The tower has been saved from demolition by the villagers, and there are plans to restore it. This was a strange, manufactured place. "With soul as long as the chapel lasts," Peter said.

Onwards to Riley Green for the night. Peter and I took it in turns to walk the towpath. We saw a deer, some beautiful fungi, and three men on a boat. "Where's Montmorency?" asked Peter. Only one of them got it.

Moored up at Riley Green, we visited Hoghton Tower, the home of the de Hoghton family since the Norman Conquest. It was there, on the hill above, just waiting for us. The house gave stunning views over the surrounding countryside towards the Wirral. Pendle Hill away to the north. Blackburn to come, in the east. The hay harvest looked perfect in the warm sunshine as we walked alongside a wall, a thing of beauty, built with real care.

It's beautiful countryside, this West Lancashire cotton country. The accents change from place to place for those with ears to hear. Walk it at the pace of a narrowboat, and you hear the rhythms, much as you do when you read the psalms slowly. The joggers and cyclists speed by; the fishers find their places to settle for the hours ahead.

Peter walks for a bit. When we walk together, I'm always out in front. "I'm worried," I say to him, "that if I slow down to your pace, you'll slow down even further, and we'll gradually stop." An interesting reflection on our marriage—he lives with his frustration that this is what I do, striding away too fast, just as I live with the anxiety that there's always more life to be had, more to give. Then, when we read the psalms together at morning prayer, I leave much longer pauses. He's more likely to cut them short and hurry on to the next verse.

Pace. Rhythm. So important. "It feels like I'm finding a different pace and rhythm to my life," I say to Peter. As he is. We're learning to walk together after years of walking differently. He has been striding the corridors of hospitals at a pace I can barely keep up. I know. I've tried. Now he wants to go too slow.

Walking is something to write about these days: lots of books written about it. I love the thought of *Solvitur ambulando*—that "It is solved by walking". It was perhaps first coined when Diogenes walked out on Zeno as he contemplated the philosophical problem of the reality of motion, thus proving that it is indeed solved by walking. Then Augustine of Hippo is meant to have written about it. Walking—like narrowboating—takes you along at three miles an hour. Slow enough to absorb the environment, fast enough to get us somewhere, eventually.

There's a great passage from the prophecy of Isaiah (30:8–21) that ends with the words:

> And when you turn to the right or when you turn to the left, your ears shall hear a word behind you, saying, "This is the way; walk in it."
>
> *Isaiah 30:21*

This is the way, walk in it

We are path-makers. It doesn't take long to create a path. It's often formed by use: folk walking it regularly, tracing the footprints of others. I give thanks for paths, and for those who create them; for the fact that they come and go over the terrain, appearing and disappearing over time as people need them. A gift for all of us, to guide our feet. Using a path is a communal activity. Of course, we could all branch out and forge our own way over the ground. And sometimes we have to—the first along a path will have done this, whether around a field on the outskirts of a town, over a hill or a mountain, or one of the long, old paths that criss-cross our land: the Icknield Way, the Coast to Coast, the Cumbria Way.

Robert Macfarlane's book *The Old Ways* explores the ancient paths over land and sea; a fascinating account of how humans interact with their natural environment, marking, and being shaped, by the land. For once a path is there, we will naturally follow it, pleased to be spared the trouble of treading down nettles and docks, avoiding trees and brambles. Glad that we don't need to concentrate completely on where our next footstep will be, but able to look up and enjoy the scenery, or study the weather. This is the way; walk in it. We have others to thank, and then we also contribute to the making of a path, its history.[156]

It's no accident that Jesus said of himself that he is the way, the truth, and the life, for he was encouraging his followers to take the way, to follow the path that he himself laid out. A path, a way, that leads to eternal life, by way of the cross, the tree of life. As his followers make their way along that road, they listen to the word which guides when there's the temptation to deviate, to the right, or to the left.

As I struggle with fear for the future, I want to be on the right path. The path I am meant to follow—for each of us has a divinely ordained way, within God's created order. We can struggle: distracted by many things, stumbling through despair, losing our way. It can be difficult to follow the injunction "This is the way; walk in it", towards the fruitful and meaningful life that is meant to be for each created being. The way of our lives, from birth to death. Accompanied by the Holy Spirit, we follow the way, the truth, and the life, Jesus Christ. Christ who guides us from God, to God.

◆ ◆ ◆

There are many now who are following the path of Greta—who are forging ahead as Extinction Rebellion. I know I'm not alone. I need to join forces and tread the path that others are following now, into activism. Perhaps the particular activism of enabling a contemplative spirit to open up the path from despair to hope, through lament.

What does this place know of me that I cannot know of myself?

One winter Peter and I did the same walk up Black Combe on two consecutive days, 11 and 12 January 2013. We were interested in whether we would see the same things, feel the same feelings. I wrote a pair of poems to describe what we experienced, trying to answer the question that Macfarlane asks: "What does this place know of me that I cannot know of myself?"[157] Perhaps that's a question we need to struggle with as we contemplate the land that cries aloud for healing. Perhaps we are crying aloud for healing too, and don't know it.

Here's the first, "Black Combe, 11 January 2013":

> We walked Black Combe from Whitbeck yesterday.
> Above the road, a newish path through bramble
> and bracken to Whicham where Wainwright's path,
> the carpet-slipper one, begins. Long rambled,
> foot trod, it takes you to the top. Or not quite;
> bypassing the trig point in haste to be gone down
> again, or off to those distant hills; the Gable, Old Man,
> Esk Hause, Scafell Pike. It pays off, this relentless route.
> Or did for us. Clear above the haar that up the Valley came,
> the sea to west, the mountains east, a flock of golden
> plovers, and silence like you never heard before.
> Christmas cake at that familiar cairn
> and before we got too cold we strode, Bootle
> Town below, sweeping down on cracking ice
> above the loamy bog until the lovely green velvet
> road, the ancient road, corralled us down
> to that distinctive field. Then left. We contoured round
> following the wall over becks and passed the old
> derelict farm that once was grand. A fell cottage.
> Hollies, old thorns, stoved-in dinghy. Intriguing,
> with its windows and its padlocked door, it seduced
> us. We wondered to whom it belonged. Would they
> sell? But we've fantasized like this before.
> The Combe from here allows you in now,
> and then as steep-cut becks sheet their white
> way down and leave one imagining what was
> the top, so secret now, Whitbeck Mill, fire embers
> smoking hot. And then a micro-climate; a garden full
> of flowers: unseasonable periwinkle, lily, wallflower,
> cyclamen, all out. Puddles, potholes underfoot, and
> Please Don't Let Your Dog Foul This Lane.
> The church, St Mary's, full of residents who once

lived in those farms and houses we've just passed.
Snowdrops. Churchyard celandine, daffodils a promise.
We walk in faith and knowledge this path today. Wondering
the difference. It traces itself in memory; does it remember
us? What do we know that can't be known elsewhere?
We shall walk this route, we said, every day, when we retire.

And the second, "Black Combe, 12 January 2013":

The following day we sat in bed and decided
to walk Black Combe again. To retrace our steps
to see if the hill remembered us. Colder, and cloud.
Both in a different mood, we argued; one wanted to take
the high route; the other to follow yesterday's path.
We did. Resistance, internal, when you know
what's ahead; anticipating the steep places, forgetting
the joy. We walked in cloud, more slowly; comparing
our tracks, matching our prints. Displaced rocks still
displaced; orange peel freshly dropped. The dog
happier, less timid. My blackthorn stick, old, familiar,
forgotten the day before. Purple leather gloves restricted
my grip causing muscle ache, upper arm. We met two men
as we emerged from the cloud. Glorious, glorious.
White waves of cloud below, we flew above, on
upward; that steep last pull; and then again the cairn;
the view: big hills like islands in a sea of white.
The last of the Christmas cake and an orange for lunch
and the cloud again. No Bootle; sheep bounding away
in the mist. We drank again at Holegill Beck, under the
sycamore trees, and paused again at that old fell
cottage. This time the fantasy stronger. What if? Should
we ask at the farm? At Monk Foss Farm? (We later gather
that no, the Wilsons don't own it; a hermit, Rigby, once lived there.)
This time we detoured above Whitbeck Mill to discover the pond.
The black hill, tickled lightly that we'd returned. I've seen
it shudder, violently, moodily; stirred by ancient memory.

◆ ◆ ◆

The footprint we leave on the planet should be at the forefront of our minds. Now is the time, with increasing urgency, that we must push for political commitment to meet those UN targets. We all need to travel that path from despair to hope through lament, and encourage others to walk with us, asking all the time, "What does this place know of me that I don't know of myself?" It's the question that takes us towards the place where we can begin to change as we must.

Blackburn

The next morning, and we're at the first lock in Blackburn by nine o'clock. "That was a bit of a bugger," says Peter, as we move on to the next; the paddles were hard to lift. But then we meet another boat coming down the flight of six, and both grin. That means all the other locks will be in our favour—empty. Lock 3 opens as if by magic, while Peter is still behind me, closing the paddles on Lock 2.

There's a lock keeper. I ask him where's best to moor. "Best to go on through Blackburn," he says, intimating that there's nowhere that's really secure. "No, we really want to stop in Blackburn for the night," I explain.

"That last boat moored up at Eanam Wharf. They had no trouble. There's nowt else." He leans on the gate as the lock fills, and the Lark ascends, slowly, surrounded by muddy water and plastic bottles. I muse on the psalms, and water.

> I waited patiently for the Lord;
> he inclined to me and heard my cry.
> He brought me out of the roaring pit, out of the mire and clay;
> he set my feet upon a rock and made my footing sure.
> He has put a new song in my mouth, a song of praise to our God;
> many shall see and fear and put their trust in the Lord.
>
> *Psalm 40:1–3*

Peter catches up and leans on the other gate. "Where's a good place to moor?" I heard him ask. I could hear, above the sound of the engine and the swirling of the water around, ten feet below, the lock keeper repeat what he'd said to me. "Eanam Wharf is just beyond Bridge 103. You won't miss it."

There's a couple of other boats there, so we moor up directly outside the Calypso Restaurant, where we meet Jonathan and Emma for a meal. We have a wonderful platter of Caribbean food laid before us, and talk of Mirfield, of the seriousness of the life there, its disciplines and routines, its communal life and prayer. We talk about the psalms, and how frustrated Jonathan was to begin with, saying them so slowly. But now, how he loves it. It's as if you chew them over, I say, tucking into jerk chicken.

Jonathan's on placement in Ribchester. They are both relaxed about where he'll serve his curacy, though Emma wonders whether she'll have to find another job. We talk about "somewhere people" and "nowhere people", and I say a little about the book *The Road to Somewhere* by David Goodhart.[158] Roots are important.

Pudding was pineapple and banana in a wonderful Jamaica Rum sauce, then we returned for tea on the boat, stepping out of the restaurant and over the fence. "I must talk to her," says a rather inebriated young woman, sitting outside the pub, pointing at me. "I've always wanted a boat like that," she slurs. "When I grow up, I'm going to get a boat like that." Her boyfriend encourages her to come along. We say goodnight to her, and that we hope she finds her dream. We descend, closing the hatch doors over our privacy.

More of Blackburn to come. There's the Cathedral. And the Museum. And a city that's filling its four thousand holes with much evidence of regeneration and vibrancy.

CHAPTER 13

Onwards to Pendle

It's three weeks to Peter's ordination, and I have been seriously panicking about his stole.

The usual thing is that this piece of ecclesiastical kit is given to the ordinand as a present, to mark the beginning of their public, ordained ministry. It derives from the towel that a servant would carry over their shoulder in classical times, and for a deacon it's worn over one shoulder and joined at the waist. When someone is ordained priest, it goes around the neck and hangs down on either side towards the knees.

When I was ordained deacon in 1989, I was given a stole made by Philip Manser. He lived in Westhoughton, where I served my curacy, and made the most beautiful embroidered clerical robes. When I was ordained priest in 1994, he made me a chasuble depicting Mary Magdalene with long red hair, and a whole set of stoles for each of the seasons—green, purple, red, white—which I still have.

Philip had moved to South Africa, and we'd lost touch over the years. I Googled him, to see if he was still making them—but couldn't find any address for him. I had it in mind that it would be good to give Peter a stole Philip had made.

When the search came to nothing, I then thought I'd make the stole myself. I had a vague idea of what the design might be, based on a stylized image of grapes that I found in a book of William Morris artwork. I set to, sewing away. But it never really worked in my mind. I kept changing the overall concept. To be honest, there *wasn't* an overall concept.

So when it came to the blessing of the stoles at Mirfield after the last Mass a week or so ago, I bought some embroidery threads and gold linen and Father Peter blessed them, to be incorporated into the eventual stole. Even that didn't help.

We arrived at Blackburn Cathedral on Friday afternoon, to meet Canon Rowena for tea at 3.30. She showed us around beforehand. It's a lovely, airy, light building, with some arresting artwork.

Blackburn Cathedral

Blackburn was carved out of the Diocese of Manchester in the 1920s. From the 1930s onwards, an extension scheme was devised to turn the former parish church into a cathedral worthy of the name, including a central tower of Gothic proportions. With funding sources compromised by the Second World War, the plan was simplified to a concrete central corona, designed by the architect Laurence King, with artwork by John Hayward. It was all completed by 1977, when the Cathedral was consecrated to serve the Diocese of Blackburn as its mother church and the seat of the Bishop's apostolic ministry. Canon Rowena's tour included the artwork that's all around.

The Passion—the story of Christ's agony, trials, suffering, and death—is vividly portrayed in a sequence of figurative Stations of the Cross, painted by Penny Warden and installed in 2005, entitled "The Journey". Mary particularly caught my eye, as did Veronica. John Hayward's "Christ the Worker" is above the West Door, suspended on a loom, a reminder of Lancashire's cotton and weaving past. There's a "Madonna with Child" by Josefina de Vasconcellos, and the "Resurrected Christ" in the Jesus Chapel was modelled on Hayward's niece. The seraphim that soar above in the lantern were also designed by John Hayward. The overall effect is lovely; feminine too, as befitting a Cathedral dedicated to St Mary the Virgin.

The best thing of all—personally—is that there in the shop were some stoles for sale. Made by Philip Manser.

Peter and I returned for Mass on Saturday morning, and Canon Rowena arranged for the shop to be opened. Bishop Philip explained how the Cathedral came by them: that a woman from Haslingden had contacted him, saying she had these stoles made by Philip Manser from ages back—might ordinands perhaps want them? It's a lovely thought that Peter will be ordained in a stole made by the same person who made mine, all those years ago. And wonderful that Bishop Philip had time to bless it too.

Blackburn Museum and Art Gallery hosts a great collection of religious icons and manuscripts, collected by Robert Edward Hart (1878–1946). The Blackburn Psalter is there, made in Oxford around 1260 by monks trained in Paris.

"Beatus vir"

Psalm 1, known as *Beatus vir*, which begins "Blessed is the man who walks not in the ways of the ungodly," has an illustrated capital B showing King David playing the harp, with the judgement of Solomon below. Ellen Davis says that it's not really a psalm at all, but a wisdom poem designed, "like any good preface", to tell us "what we need to know from the outset in order to make sense of the rest of the book".[159] The most important thing is to read God's Word, all the time, she says.

Those psalms, which Peter and I say together as we pray every morning, have shaped people of faith in the Judaeo-Christian tradition since time immemorial. It's a blessing to be shaped in that tradition today.

> Blessed are they who have not walked
> in the counsel of the wicked,
> nor lingered in the way of sinners,
> nor sat in the assembly of the scornful.
> Their delight is in the law of the Lord
> and they meditate on his law day and night.
> Like a tree planted by streams of water
> bearing fruit in due season, with leaves that do not wither,
> whatever they do, it shall prosper.
>
> *Psalm 1:1–3*

◆ ◆ ◆

We leave Blackburn accompanied by Lois, who will be ordained at Blackburn Cathedral next year. She is training at Mirfield, and we talked

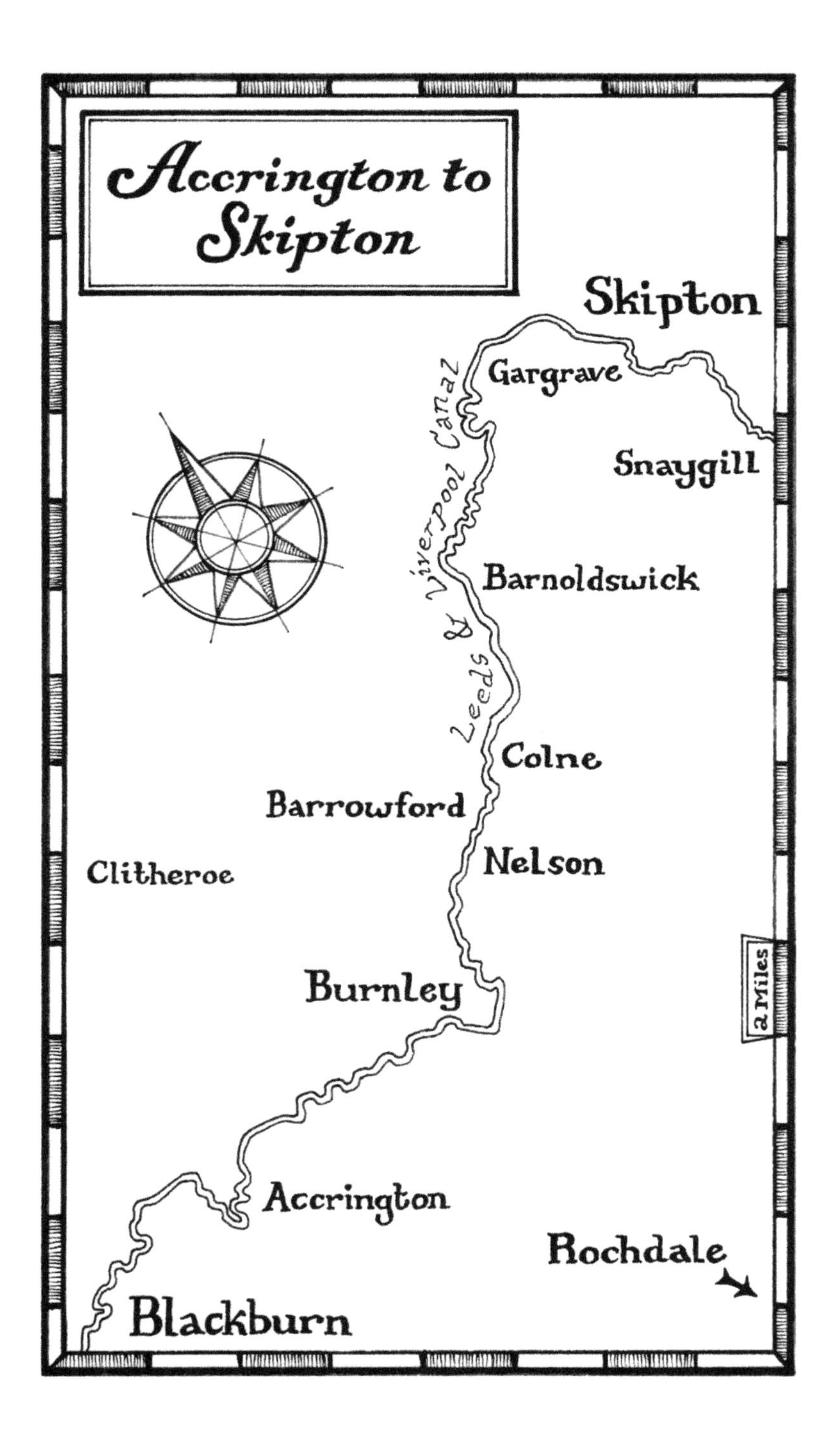
Accrington to Skipton
Skipton
Gargrave
Leeds & Liverpool Canal
Snaygill
Barnoldswick
Colne
Barrowford
Clitheroe
Nelson
2 Miles
Burnley
Accrington
Rochdale
Blackburn

of her excitement at the prospect of ministry in Lancashire. She'll be a great priest.

Spending time with ordinands has put me in touch with my own sense of vocation again. I'm chewing over the question of what God wants of me for the next decade of my life. My passions—anxieties, desires, fears, hopes—help in discerning. I mull over the advice I was given by Bishop Peter Walker, so long ago, about being called to be a priest, a pastor, a preacher. Now, the prophet and the poet stir within me. What should the Church offer to serve the world, instead of being so preoccupied by its own decline and working so hard for growth? That's what this voyage/pilgrimage is all about. Are we brave enough to lament this passing age with fierce hope, contemplating the beauty and the tragedy of what lies ahead? What more could I do, to bring experience and gifts to that overwhelming question of how we cope with the possibility of a catastrophic future?

"Welcome to Church"—Accrington

When we lived in Westhoughton, our home was on a housing estate built in the post-war period. It was built—as was the whole estate—of Accrington brick. It's a bright, hard, red brick that gives little to the eye, and even less if you want to drill a hole or two in it. That and the name of the football club—Accrington Stanley—was all I knew about Accrington. "Who was Stanley?" asked Peter.

We enjoyed the excitement of the aqueduct over the M65, and then moored up at Church, on the outskirts of Accrington, having appreciated the Canal and River Trust sign which read "Welcome to Church". Nice to see a secular organization welcoming folk in such a way. But the church in Church is sad, very sad. Derelict.

The mooring wasn't so hot either—given that this is exactly halfway between Liverpool and Leeds. There's a sculpture to mark the spot. We litter-picked the stretch. And then we wandered into town, through terraced streets and tired-looking shops of Asian clothing and fabrics, with a great, gleaming mosque in sight from the Blackburn Road. It's the

£8.6 million super-mosque, the cause of much controversy both before and after it opened its doors in December 2017.

We walk down to the viaduct that carries the railway line, and then up again, through terrace after terrace, following my phone's directions. One street was full of St George flags. "For the World Cup," said Peter. I said they looked like they'd been up since the Brexit referendum.

We were looking for the Church of St Mary Magdalen to check the times of services for Sunday. They have a website, but that insists we join Facebook, which I don't want to do. I call the number on the website "A Church Near You" and leave a message. Father Lawrence gets back to me later. We say we'll be at the Solemn Mass at eleven in the morning.

The church is next door to a Church of England Aided primary school in the middle of steep terraces that run in parallel down to the town. This is a largely white area, it appears. We walk past a couple of children outside their front door, as another inside calls out, "Is that Mum come home?" Perhaps she was the tired woman who we ask for directions—who looks like she holds down two or three jobs and has little energy for what waits for her at home. Peter surmised her husband, or partner, would be unemployed and depressed and not much use at all. If he's still around. He'd seen a lot of families struggling like that in Bolton. Really tough, this hopeless, depressed poverty.

On Sunday morning before church, Peter meets Dan who was in the army and now works at Travis Perkins. "I used to fish here when I was a kid," he says, as Peter picks litter around him. He loves wildlife, he says; often takes his nephew and nieces to learn the different trees, birds, and flowers. Most weekends he heads off down the towpath, rambles along the canal. "My wife doesn't come; she doesn't want to. I'm happy by myself." Broad Accrington accent, hard to capture in print. He was thrilled that Peter was collecting the rubbish.

Towpaths are used more than you'd think. Joggers, cyclists, of course; but also families out walking. One woman and her two small boys walk all the way to Rishton and back—we saw them go and return. Took nearly two hours. Stunning views out over the Calder Valley and towards the Pennines to the north.

Accrington Brick is often called NORI—why?

As we came towards Accrington, we saw strange ovens or kilns alongside the canal and wondered what they were. I thought they might have been old brick kilns. Googling it later, we find out they are to cook coke. The bricks are produced on an altogether bigger scale. Accrington brick is often called NORI—why? Because the letters IRON were accidentally placed backwards in the brick moulds, thus spelling NORI. Well—that's the story.

Bricks were produced here, at Altham near Accrington, from 1887 to 2008 and—good to know—again from 2015, in response to the need for building materials. The bricks are famed for their strength. They were used for the foundations of the Blackpool Tower and the Empire State Building. The bricks are also acid-resistant, so are good for lining flues and chimneys too.

Why here? At the end of the Ice Age, the River Calder was blocked and formed a large lake in the Accrington area. The sediment from this lake produced the fireclay seams, and local coal was available to fire it. There were four brickyards originally, producing engineering bricks (Enfields, Whinney Hills) and specials. Specials are hand-thrown into plaster of Paris moulds. They can be extremely decorative.

The site has its own mineral railway connecting with the East Lancashire Line at Huncoat Station, and was once managed by Marshall Clay Products, then bought out by Hanson, a subsidiary of the multinational Heidelberg Cement Group, in 2005. When the 2008 recession hit, the factory was mothballed with the loss of eighty-three jobs.

Following an upturn in housebuilding, Hanson reopened the Accrington factory, with production starting again in January 2015. The factory has capacity to produce forty-five million bricks a year from the adjoining quarry, which has between thirty and forty years of clay reserves. The bricks are made by crushing the clay and mixing with water, to mould into different shapes and sizes, and then firing at a very high temperature to make the product hard and weather-resistant. But, I asked myself, what of building—for the future—houses that are as energy-efficient as possible? Industries like this should already be there, producing state-of-the-art materials that mean we can turn our heating right down, even off. The technology is there already. Why, oh why are we so behind? Where is the political will to implement the future we must have if we are to survive?

The service was serious religion, taken seriously

Sunday morning came, and we made it in plenty of time for church at St Mary Magdalen, finding a faster route through a much newer housing estate. New housing is going up all over, which is good for the brickworks. I hum Joni Mitchell's song "Big Yellow Taxi". They put up a parking lot.

It was the walk of witness that afternoon, we were told, explaining the emptiness. "We'll just have to move around a lot," said John, in the pew in front of us, as he caught up on the news with three women in front of him. The choir of four women sang lustily, holding the tune in the absence of the organist. "He's got the lurgy," explained the vicar at the beginning of the liturgy.

No fresh messiness here. But no youngsters either.

As we left, Father Lawrence was thrilled to hear that Peter had trained at Mirfield. I'm not sure he'd have been as thrilled had he known I was ordained. "That's a pretty dress," one of the women said to me.

◆ ◆ ◆

Back on *The Lark Ascending*, and we carry onwards, as the canal contours around fields along the M65, past Junction 8 which was our turning as we drove home from the Lakes, down the A56 to Bury, when we lived there. "Do you remember we saw a boat when we were driving along the motorway? How surprised we were there was a canal so high up?" asked Peter. "Who'd have thought we'd be here now." "I reckon that car there is saying exactly the same about us," I conjectured.

Past derelict farmhouses, and through swing bridges—tough to open, heavy to get to swing, and (once swinging) hard to stop—we made our ways onwards, slowly, to Hapton. "Best stop on this mooring," said Jed (the bloke we'd first met at Botany Bay, giving us a lift into Chorley, recommending potatoes), as he chugged past. He'd been at a party the night before in Blackburn. "There's gypsies just around the corner." We had already moored. We were tempted to move and find that further mooring. Traveller folk get a hard deal.

Instead we wandered into town, over the motorway to find the station, and then stopped at The Railway for a pint of Theakston's on the way home.

◆ ◆ ◆

Foulridge Tunnel is traffic-lighted. We had a green light in our favour as we approached and went in, following the hire boat we'd paired up with through the Barrowford Locks. It had been the second tunnel we'd done in two days as higher and higher we'd risen, passing through Gannow Tunnel as we left Hapton and came towards Burnley. Through Burnley and Nelson the terrain begins to feel like high country—because it is. Foulridge, the second tunnel, doesn't mark the border between Lancashire and Yorkshire (that comes later, after Barnoldswick), but it felt as if it did.

Singing Kiwis

This hire boat, from Silsden, near Keighley, was full of Kiwis: six of them. They were three couples, all in their seventies and eighties, who had known each other for over forty years. The boat was no bigger than ours—so they were hugger-mugger. The blokes—Jack, Malcolm, and Terry—had sung together at national and international level, winning awards in a barbershop quartet. Babs, Carrie, and Sarah—"They decide everything," Terry told me.

They'd done this before—come over for a narrowboat holiday. It'd been the Avon Ring last time. Jack, Terry, and I, on the boats (with the others managing locks), talked of Christchurch after the earthquake, and how the city couldn't decide what to rebuild, and what to leave. We talked of how Jack went out to New Zealand with his mum and dad, who were following his older brother and sister, who'd emigrated after World War Two and made their life there. Residency was easy then. Now they come back to the UK often, when it's winter there, summer here.

As soon as we were in the tunnel they chimed up with "In Dublin's fair city". The sound echoed around, filling the length of the almost-mile-long tunnel. "Now her ghost wheels her barrow, through streets broad and narrow, crying, 'Cockles and mussels, alive, alive-o.'" Our turn.

Peter and I gave them "I'll sing you one-oh! Green grow the rushes-oh." All the verses, so plenty of opportunity to give it all they had on "Three, three, the rivals." I *mean* all they had! The rounded walls that dripped and ran with water and salts had a great acoustic. The extended harmonies bounced off water and reverberated, embracing the boat that came behind.

The Summit of the Leeds and Liverpool

They were off down towards Leeds as we were, for Foulridge Tunnel marks the Summit of the Leeds and Liverpool. We are at the watershed, on the top of surrounding country, so the flow will carry us as we make our way down. The locks will empty with us, the lark descending through Barnoldswick, Gargrave, and to our final destination, Skipton.

"On Ilkley Moor baht'at" came next: all verses. "Then t'worms'll cum and eat thee oop." "Then we shall go an' ate oop ducks," we sang. As the light at the end grew greater, "Should auld acquaintance be forgot" was their goodbye.

◆ ◆ ◆

The idea of a tunnel between Leeds and Preston was first conceived in 1765 by John Stanhope of Calverley, near Leeds. A public meeting was held in Bradford the following year, and by 1767 the proposal had evolved to the plan for a canal from Leeds to Liverpool. In 1768 the Lancashire side met, and Liverpool promoters suggested that the canal should pass through Burnley and Blackburn instead of via Whalley. 1770 saw the first Leeds and Liverpool Act passed, authorizing a line via Skipton, Gargrave, Colne, Whalley, Walton-le-Dale, and Parbold.

In 1773 the canal opened from Bingley to Skipton, and the following year from Liverpool to Gathurst, then by the Douglas Navigation to Wigan. Skipton to Gargrave opened, and Bradford to Shipley and Bingley. In 1777 the canal from Skipton to Leeds was completed, but all the available capital ran out, so construction ceased on the main line. Meanwhile the canal opened from Gathurst to Wigan in 1780, and the branch canal from Burscough to Rufford and Sollom Lock in 1781.

A second Leeds and Liverpool Canal Act was passed in 1783, to raise more money, and a third in 1790, authorizing the line to be altered to avoid an aqueduct at Whalley Nab. In 1791 building recommenced west of Gargrave, and a fourth Leeds and Liverpool Act was passed in 1794, permitting a new line through East Lancashire. In 1796 the Foulridge Tunnel was completed, allowing the canal to reach to Burnley. In 1799 the Lancaster Canal stretched from Haigh to Wheelton; in 1801, the Leeds and Liverpool from Burnley to Henfield, and in 1810 on to Blackburn. In 1816 the canal was completed and opened throughout, with the Leigh Branch connecting to the Bridgewater in 1820.

Foulridge

We moor up outside the café in Foulridge, after a long day in the tunnel, and before that, the locks through Barrowford. Café Cargo has been booked for tonight's meal, and so we walk back over the route of the tunnel, looking for the three air shafts that illuminated us briefly with light from above as we made our way through. We find two of them, with brick surrounds, on the old towpath which is now a cycle route. The cycle routes that criss-cross the country, managed by the charity Sustrans, are impressive. The old route for the horses, as the boats went through the tunnel, is now part of Route 68. Peter and I find orange hawkweed in the churchyard of St Michael and All Angels, Foulridge.

◆ ◆ ◆

Back on *The Lark Ascending*, and we settle down to watch the day cruiser moored behind us get ready for a coachload from Huddersfield. The care and patience are commendable, enabling a number of older folk, with their mobility walkers, to negotiate the flagstones of the wharf, and make their way down onto the *Marton Emperor*. Off she goes, when they are all seated at their tables, down towards Barnoldswick. "Someone saw an otter yesterday," says Martin, the owner of the boat. He'd wondered if it were a mink, but no, "It was too big and brown for that."

Now I don't have to work at Peter's stole—and I've embroidered the Christogram from Mirfield onto its reverse side, using the threads that were blessed at the last Mass—I've turned my hand to a rag rug for the boat. I sit and tug strips of rag through hessian, attracting attention from the old ladies. "Good to see that's still happening," says one.

◆ ◆ ◆

Peter reads aloud from Iain McGilchrist's *The Master and his Emissary*, chapters about the Enlightenment and the beginning of Romanticism. McGilchrist explores how, at different times, Western culture has been dominated by either the left hemisphere or the right hemisphere of the human brain. He's very clear that this is a metaphorical way of reading cultural history. It makes a lot of sense to Peter and me.

When the left hemisphere is dominant, the attention becomes focused on detail and systems. It seeks to control and categorize, to differentiate and manage. When the right hemisphere is in ascendancy, the mind is more aware of that which is beyond, and other, and which cannot be grasped but only encountered. It's a great book, offering a ground-breaking way of seeing the human condition.

◆ ◆ ◆

It feels strange to leave Lancashire behind. We looked back over the last few days of steady climb from Hapton and Accrington. The view of Pendle Hill had always been with us. We hadn't stopped in Burnley—though we were impressed by the signs of regeneration, especially of the wonderful great mills and warehouses that line this canal, reminders of the industrial past. Burnley football ground had been on our left, and then the bus station painted in claret. As it made its way through the town, the canal gave us views of large churches surrounded by terraces of houses.

This was a canal with too many shopping trolleys, debris, even derelict boats; too risky to stop. As we had filled up the drinking water tank, there was a boat with a bed base attached to his stern. "It's all around my prop," said the man on board. "It's burned out the engine and the gear box. It's too tangled to remove. We're not going anywhere," he told us as we slowed down past him. What would you do, we wondered, very glad that wasn't our fate. Tuesday evening we'd moored just beyond Reedley Marina. The towpath was beautiful and used by lots of people. One man of Asian heritage was delighted that we were growing coriander on the roof.

Now, moored up at Foulridge early on Thursday morning, there's a gale blowing outside. The trees are wild, the boat pulling against its lines. We've been blessed with so little wind over the last six weeks, it's strange to feel its force.

This is Pendle country, here where Lancashire becomes Yorkshire. It would have been remote in the seventeenth century. Awful roads, where they existed; no canals. A strange and different spirit in the air.

Pendle country

That whole era of the so-called dawn of Enlightenment had far-reaching social, religious, and psychological impact. John Buchan captures it brilliantly in his novel *Witch Wood*, set in remote Scotland.[160] How the old Catholic order that seemed so natural gave way to the forensic puritanism of the new religion, which didn't know what to do with what it didn't understand and couldn't control. The right hemisphere, taken over by the left. The "uncanny" is born at that moment, says McGilchrist, drawing on the work of Terry Castle.[161]

McGilchrist explains how Castle looks back to Freud's essay of 1919 on "The 'Uncanny'" as she explores the phantasmagoria, grotesquerie, carnivalesque travesty, hallucinatory reveries, paranoia, and nightmarish fantasy that accompanied so-called Enlightenment. McGilchrist believes that the Enlightenment, dominated by left-hemisphere rationalism and culture, had no room for the magical, the wild side, the poetic, the strange.[162] In such enlightened, pure times, old rituals became suspect and strange. All around here, in this wild land with its brooding hills, witches were uncanny, beyond the ken, hunted down.

I get restless when things are too forensic. I begin to wonder what's being lost. What of our participation in the whole that is other to us, and which passes our human understanding? The Other that we encounter, rather than control. Today, in another time of left-hemisphere dominance, so much is being lost it feels unbearable. I live with a constant restlessness for the loss of God, of the deeper knowledge that transcends our shallow grasp of things. The ultimate being—the "Ground of Being", as Paul Tillich called it—whose substance under-stands all things. Now, without that glorious unity, all aspects of life lose reverence, lose their sacredness. We end up instrumental in our approach, so everything, even our attention, is susceptible to utility—relationships, the world around, all things, and all being. Everything can become commodified by neo-liberal markets. The abundance and biodiversity become ours to use and abuse, to exploit beyond endurance, until only a few dominant species are left, and then nothing ultimately survives the destruction of food chains. We face catastrophic insect collapse. That means, also, everything that depends on them.

These canals speak of a life beyond commerce. Now useless, they still have their locks and tunnels, their narrow waterways that burgeon all around with wildlife, with dereliction and untidiness. Now they are places where loss is held, as the past becomes the present. They offer the opportunity to discover a wilder side to life, to engage the right hemisphere with attention to that which is forgotten, beyond our ken, abjected, and lost. To lament is to engage the right hemisphere. It is to cry aloud in pain and distress for what is in danger, for what is lost. Grief pours out its misery, and when the grief is for the future we are caught in a double bind. For normally we grieve for what is past, and we hope for what is to come. The lament we now need to learn takes us in uncharted directions; we can't do it without God, the ultimate other of right-hemisphere attention.

> Out of the depths I cry to you, O Lord.

CHAPTER 14

A Time of Deep Engagement

Thursday 14 June was the first day of real wind, as Storm Hector blew across the British Isles. We'd read in the news that the Operator Manager for Transport Scotland had said, "The strong winds and rain may lead to difficult driving conditions, particularly for high-sided vehicles. As always, motorists should take extra time to plan journeys, follow police advice and drive appropriate to conditions. The strong wind may impact rail, air and ferry services, so travellers should check with operators to see if their journeys will be affected."

Nothing about narrowboats there. They are surprisingly susceptible. You'd think their weight and power would simply plough on through any wind—but it's easy to be blown off course and end up on a lee bank. Then it's very hard, almost impossible, to manoeuvre off, with the wind holding you there.

Thursday dawned with the gale still raging, and we set off from Foulridge northwards towards Barnoldswick—Barlick, as the locals call it—and from there, towards Gargrave, where we thought we'd stay the night. If we got there.

The Met Office has started naming storms: it helps to alert people that they need to be ready for this one. And Hector was wild—with the trees thrashing around, threatening to drop branches on us as we drove along. Thankfully not many other boats, as we had one or two near misses, the wind pushing us off course towards moored vessels.

This stretch of the canal is beautiful—perhaps the most beautiful of the whole Leeds and Liverpool—so we'll have to come back someday, to enjoy it without the strain and concentration required by Hector. The countryside is green and rolling, with the canal contouring around through fields and villages.

Are we still in Lancashire? It doesn't feel like it any more. It feels like Yorkshire. Pendle Hill is still brooding over us to the rear, but very soon the Yorkshire Moors appear in the distance before us. The Pennine Way joins the canal towpath for stretches, on its way to the Scottish Borders.[163]

We pull up in Barnoldswick

A bloke with his dog stops to talk, and to commend the town. The Rolls-Royce social club is the place to eat, should we be here overnight. There is a real, traditional grocer's in town. The dog—a Staffy (true, not American, with a lovely smile and boisterousness)—falls in the canal as she misjudges the boat's distance. He warns her of a shower when they get home. She knows the word.

The town is obviously competing for the "Town in Bloom" award. Flowers everywhere, local businesses supporting "Barlick in Bloom"—because, of course, the name is shortened even more than Oswaldtwistle is to Ossletwistle. The town hall has it as Barnoldswick Civic Hall, the local Indian more colloquially: *Barlick Raj Balti.*

We wander, looking for the grocer's, and find some delightful shops and a market square, thriving with life. The grocer's is run by Sikhs, with a range of vegetables—some you'd expect, others not. Then a coffee at a cake shop. On our way back to *The Lark Ascending* we pass the Rolls-Royce works, just on the day when cuts of 4,600 jobs are announced. One of the reasons Barlick feels so resilient is the number of businesses we come across.

"A high, wild note of wind"

Back to the boat, and the canal has white horses. Never seen that before. The wind is hectoring—a good name. It leaves me tired, disgruntled, irritable. I think of that poem "Wedding Wind" by Larkin, of the wind of disappointment through the wedding joy of a new bride.[164] And also, more positively, of L. M. Montgomery's character Emily, who would imagine the Wind Woman bringing her poetic soul to life, in her *Emily of New Moon*:

> It had always seemed to Emily, ever since she could remember, that she was very, very near to a world of wonderful beauty. Between it and herself hung only a thin curtain; she could never draw the curtain aside—but sometimes, just for a moment, a wind fluttered it, and then it was as if she caught a glimpse of the enchanting realm beyond—only a glimpse—and heard a note of unearthly music.
>
> This moment came rarely—went swiftly, leaving her breathless with the inexpressible delight of it. She could never recall it—never summon it—never pretend it; but the wonder of it stayed with her for days. It never came twice with the same thing. Tonight the dark boughs against that far-off sky had given it. It had come with a high, wild note of wind in the night, with a shadow-wave over a ripe field, with a greybird lighting on her window-sill in a storm, with the singing of "Holy, holy, holy," in church, with a glimpse of the kitchen fire when she had come home on a dark autumn night, with the spirit-like blue of ice palms on a twilit pane, with a felicitous new word when she was writing down a "description" of something. And always when the flash came to her Emily felt that life was a wonderful, mysterious thing of persistent beauty.[165]

I try to persuade myself this wind is exhilarating; that it opens up other worlds, worlds real and close to God's mysterious beauty.

◆ ◆ ◆

The Lark Ascending is ready to battle on. The wind doesn't feel on our side. After Barlick there are locks—the Greenberfield three—and then a double-arched bridge at East Marton, followed by the locks descending into Gargrave. We had help with these from CRT volunteers, who are always keen to talk. It's dairy country, and it's good to see cows (here in a dairy herd, with a bull for added measure) out grazing as cows should.

The volunteer had a view. He didn't rate the local farmers—the ones that weren't organic. "They take a lot of water out of the canal, for a start," he said. "There's a farmer locally who keeps 900 cows in his barn. They never see the grass. The grass is cut into silage and taken to them in the barns. Then the shit they produce is sprayed all over the grass, to make more grass. It doesn't make sense. The cows never see the grass," he repeated.

Many of the bridges along the canal have rope marks. Rather beautiful—not good for the ropes, though, and not good for the bridges either. So rollers were used. Most of them have gone now, but along this stretch we see one or two. And even one around a bend in the canal, to prevent the rope tangling in the hedge.

Hector means—I fear—the end of the lovely settled weather that we've enjoyed all trip.

I'm reading *Mr Lear* by Jenny Uglow at the moment. She describes how Edward Lear visited the Lake District in 1836, when he was twenty-four. He was already a fine painter of birds; perhaps he could become a landscape artist. A trip to the Lakes was a must, then, for any aspiring artist. But it rained. And rained! I must read more of Alexandra Harris on weather. In the blurb to market *Weatherland* she says it's "an exploration of imaginative responses to the weather in England" across the centuries. "I wanted to lie on the grass and watch the sky with Chaucer, with Milton, with Turner." I'll lie with her. The weather is so moody, so powerful an aspect of living in this country and countryside.

The patron goddesses of idle fellows

She reminds me of the association between clouds and minds—how bubbles are cloud-shaped in cartoons; that there was a strand in early Christian mythology that Adam's thoughts were made of cloud:

> The image is splendidly vivid: God takes a handful of cloud and shapes it into thoughts—except that, being cloud, the thoughts keep changing shape.[166]

The association continues, with dark clouds liable to press down upon us and bright, high skies without any clouds that lift the spirits—the metaphors that tie our mental states to the clouds (or lack of them) are all around.

After so long enjoying blue skies and sun, it was strange to see clouds, rain-producing clouds. Peter reminded me of the reflection from Aristophanes that clouds are the patron goddesses of idle fellows. As Coleridge wrote, in his sonnet "Fancy *In Nubibus*, or The Poet in the Clouds":

> O! it is pleasant with a heart at ease,
> Just after sunset, or by moonlight skies,
> To make the shifting clouds be what you please.[167]

Alexandra Harris wrote of how Virginia Woolf captured the bright new world of modernism in her novel *Orlando*:

> As she comes up to date, speeding towards the present time of the novel, she is aware, more than anything, of light. "The light went on becoming brighter and brighter, and she saw everything more and more clearly", until she discovers, with the striking of a clock, that it is ten in the morning on 11 October 1928. This, the "present moment", is publication day for Orlando.[168]

The bright, brittle, febrile light of today. I think of the Church's relationship with light and darkness and know why I love the service of Tenebrae so much. Traditionally held on the three days preceding Easter, candles are extinguished one by one, with the "*strepitus*" ("loud noise") sounding out in the total darkness at the end.

Psalm 139:12 captures it, in the traditional King James Authorized Version:

> Yea, the darkness hideth not from thee; but
> the night shineth as the day:
> the darkness and the light are both alike to thee.

That left hemisphere—the Enlightened, modern mind—shuns the shadows, the dark depths of God. Now, after the bright glorious May we had enjoyed, rain buffeted us from dark clouds that that seemed to go before us, holding the mystery of life, of God. The cloud of unknowing, the cloud of entanglement of all there is.

We moor that night, Thursday, on the aqueduct over the river Aire. Leaning on the parapet, with pints of beer in hand, we see a dipper below. The dipper (*Cinclus cinclus*) is black/brown, with a white breast, and is like a great water wren. Dippers feed on caddis flies and other invertebrates and insects, and we watch this one, bobbing and curtseying its way in and out of the water, searching for food. They are as much at home underwater as on the rocks, able to walk on the riverbed against the flow of the current. This bird took off, after we'd watched for some time, as if it were breaking the tension it held. I dug out a poem I'd written once, and read it to Peter—"Surface Tension":

> The air is heavy. Still. A sacrament
> with word and gesture slow, deliberate.
> We meet, and water is wine, a long moment
> of talk and play; contained, so intimate.
> The beck flows fast, the dipper turns with slow
> wingbeat in the heavier element.
> It breaks surface, powered from below;
> from water to air, in each, impatient.
> It does not pause or change its rapid flight.
> With sharp "zik" note, it whirs down stream
> through lighter air. To us, it's lost to sight
> where present tension punctures future dream.
> All forces balance. The meniscus holds.
> As heavy as honey is sweet, time unfolds.

If only.

Snaygill

We awake to the sound of curlews. It's an evocative cry, or more of a whistle, drawn out and melancholic. "A grief of curlews" was the collective noun I'd given, in that little poem "A Gospel of Birds". Good to hear that plaintive, evocative sound.

The bird life is good here at Snaygill, even with the A65 just there, on the other side of the canal and hedge. Two pairs of swans with three cygnets each: one pair are good parents, the other much less experienced (we've gathered from Janet who owns another boat). Lapwings in the fields; and yes, larks as well. We hear warblers churring away. Yesterday evening there were a good number of swifts—at least thirty—wheeling high above, and swallows, fast and skilful, as we walked the towpath to Bradley, the next village.

Our mooring is alongside another boat which is painted in almost the same colours. We're looking forward to meeting its owners. One of their friends popped along to have a chat on Friday evening. Once we get to know folk, we think we'll be happy here. It's almost exactly one hundred miles from Liverpool.

◆ ◆ ◆

We had set off from Gargrave on Friday morning early, the wind still significant, but nothing like Hector the day before. Six locks. The last one of the entire trip is the 219th that I've gone through, with Viv, or Jenny, or Peter's help. Three swing bridges too, and we are into Skipton, with three more between Skipton and Snaygill, lifting the roads, making cars and people wait.

There was Keith and Alison's boat, moored up in the middle of Skipton. We hadn't seen them since we went up the Wigan Flight with them. They weren't on board. Lots of other boats. A great atmosphere. We passed the junction of the Springs Branch, which is restricted to boats under thirty-five feet, unless you're confident you can reverse out avoiding moored craft.

The branch was opened in 1797 to enable Lord Thanet, who lived in Skipton Castle, to load his limestone, brought from the quarries by tramroad, to be taken to Leeds. It runs for half a mile, then becomes a ravine through ravishingly beautiful old woodland, now managed by the Woodland Trust. Peter and I walk up later that afternoon, after enjoying pies and mushy peas. Then we visit Holy Trinity Church. The church is lovely; we'll go back on Sunday morning.

We'd walked into Skipton from Snaygill, once we'd met the marina's owner and moored up. We'd signed a contract to say we wouldn't live aboard or have too many cars in the car park. He'd directed us to a laundry, tucked away behind the Plaza Cinema, which saw us lugging heavy bags up and down steep streets until we found it; they managed to wash and dry everything in a couple of hours. This gave us time to wander. The Oxfam bookshop was a great place: I managed to find—at last—a copy of *Lark Rise to Candleford.* A small hardback, *The World's Classics* edition.

It's been in my mind to read it again. Horribly sentimental in the TV adaptation, but an interesting piece of social history. The introduction by Hugh Massingham is well worth reading for his tripartite division of social and economic strata—now long gone, but still yearned for, I suspect (at least the romantic aspects), amongst those who voted for Brexit:

> It all seems a placid water-colour of the English school, delicately and reticently painted in and charmed by the character of Laura herself. But it is not. What Flora Thompson depicts is the utter ruin of a closely knit organic society with a richly interwoven and traditional culture that had defied every change, every aggression, except the one that established the modern world. It is notable that, though husbandry itself plays little part in the trilogy, it is the story of the irreparable calamity of the English fields. In the shell of her concealed art we hear the thunder of an ocean of change, a change tragic indeed, since nothing has taken and nothing can take the place of what has gone.[169]

And importantly, *Lark Rise to Candleford* gives a baseline for biodiversity, as it was in the 1880s.

George Monbiot, in his book *Feral*, explains that the natural world we grow up with is the one we think is normal.[170] That's the baseline we use to decide whether the natural world around us is degrading or improving. So the baseline Peter and I have is from the 1970s, before farming intensification really set in. Our memory is of species—flycatchers, hawfinches, corn buntings, butterflies—many of which are now rare, or have faded out.

Our children, on the other hand, have no such memory. What they think of as normal is ash dieback, no elm trees, and exotics like Japanese knotweed all around us. To read Flora Thompson gives another baseline—hers, from the 1880s: "[T]he white tails of rabbits bobbed in and out of the hedgerows; stoats crossed the road in front of the children's feet—swift, silent, stealthy creatures which made them shudder; . . . Bands of little blue butterflies flitted here and there or poised themselves with quivering wings on the long grass bents; bees hummed in the white clover blooms, and over all a deep silence brooded."[171]

Lark Rise to Candleford.

◆ ◆ ◆

Larkrise to Skipton.

I've been on a journey, as I've traced the daily, onward chugging from the River Lark at Prickwillow to Skipton, up through the heart of England. This was a route that isn't the normal one, any more, for getting from A to B. Six weeks, and blessed by the weather. It's been a God-given chance to think deeply about my life, and sense of vocation, as Peter and I begin a new life in Cumbria.

The canals are a delight—and should be a national park in their own right, dedicated to preserving and developing the biodiversity that flourishes in these corridors of wildlife. From them you see a different country, often high above or contouring around the local landscape. Or you're taken directly into the heart of cities, unseen by the traffic and busyness around.

They were built for one purpose, and now have another—which is to provide the opportunity to slow down—whether on a boat, or cycling, or on foot. The water and the towpaths are extraordinary. They give the chance for a different kind of attention to the world around to grow and be nurtured. An attention that engages with the right hemisphere of the brain, seeking the whole picture, the unity that permeates and inspires all things, rather than the narrow focus of an enlightened left hemisphere, in which the desire for process and control, strategy, and utility takes dominant place.

Reminding the world of the rumour of God?

How big is our whole picture, I ask myself, in today's world? How impoverished is our imagination, even without us knowing it? What of the Church, and how it should remind the world of the rumour of God? The Church, that should be ever alert to the grace that breaks through? As humanity faces up to end times, what does it mean to talk of God? Of God's love that transcends death and holds out hope that there is always something more, something beyond this life, that the human cannot grasp, cannot control? Something good and loving, a personal energy of love that holds all things and urges them on to fulfil what each is meant to be? An eternal promise of love and life that stretches all creation to yearn for consummation? A love that holds us through the judgement we bring on ourselves; through our lament for the dying of the light, the dying of the myriad complexity and wonder of the created order, so comprehensively destroyed by human greed and exploitation?

Lament is the best thing to do with despair

As I've travelled with Viv, then Jenny, then Peter, I've carried with me a continuing sense of loss and lament; with Mark Cocker, a sense of the profound losses sustained by the natural world around, reminding myself—frequently—that lament is the best thing to do with despair. We face the loss of the natural world around us, natural capital without which we can't survive. Loss of an England that is passing—many Englands that have passed—from the seventeenth century, when the Fens were drained, to the eighteenth century, when the canals were built, through their heyday in the

1830s, and then the development of different travel and communications systems, to the world of speed and instantaneity that we rely on today. We live too fast and bright, with a febrile energy that will burn itself out.

Richard Rohr commends learning to "breathe under water".[172] A strange notion, that takes us into an element where we're not at home; where we have to rely not on ourselves, but on God. We breathe under water when we enter into clouds, and there we might find God: the cloud of unknowing, where our confusion, sadness, and lament are all we have to offer.

Looking back to the past, to the construction of the world around—the canals that hold in the water, the fields enclosed, or drained and claimed from peat—nothing is untouched by humanity. This voyage was a lifelong opportunity to attend to the Other, all around, the deeper meaning that lies beneath the surface, and to attend to the deep yearnings of life. I know my own deeply felt yearning for an England that is past, and its different baselines through the ages; wishing myself back to a childhood time where anxiety—at least, *this* anxiety—was unknown; a childhood of slow immersion in the mystical numinous, participation in the life that throbs through the smallest atom, the wind, the storm clouds, the flowers, and wildlife around. I find myself wanting to be there, instead of in this life, which is too fast, too crowded, too lonely.

I've looked back to the past—my own, and this country's—with an attention to detail that tries to capture the weft and warp of the fabric of existence, that takes me deeper, beyond a shallow attention. To find resources which can answer the anxiety that destroys hope. As I've contemplated my own anxieties about the future, in a world where human social collapse, even extinction, is a serious threat, I've turned to the rich resources of the culture, nature, literature of a civilization that has been painstakingly built, and that, somehow, the canals of England represent, as we journeyed into the past, and through the present.

A re-engagement with the natural world around, with the emotional and moral knowledge of a rich cultural past, enables us to flourish more fully as human beings. It enables a generosity and openness of spirit that the abundance of nature and the treasures of poetry and literature can inspire. We need better to frame that yearning as a desire for belonging in a world where humanity lives in the natural world—God's creation—with a greater sense of reverence and holiness.

Sanctus, Sanctus, Sanctus

And the Church, trying to grow in a world that doesn't care any more for God, works best, perhaps, when it offers belonging as rootedness: of living alongside others; of caring for the stranger and neighbour. The Church doesn't thrive in a world where all is instrumental and contractual. And it hasn't found a way to commend its traditions to generations who have not been formed with church-going, choir-singing habits.

The churches we've visited have held a sense of the Other, of God—Ely, Wadenhoe, Peterborough, Brinklow, Rugby, Polesworth, Manchester, Wigan, Blackburn, Accrington, Skipton—and so many more. When we give these wonderful places the attention they deserve, allowing the right hemisphere to soak in the encounter, we are in touch with a world beyond. The Church is not there merely to enable our narrow ideas of mission and growth (even when it's so often seen as an impediment to that mission). It's there to enable us to engage our right-hemisphere attention, and close down the left hemisphere's dominant desire to analyse, to control, to process. The Church is there to remind us that grace breaks through, always and everywhere.

That's what I've been seeking as I've undertaken this personal voyage: my own re-engagement with what is Other to me—in the natural world around; in the different element of water; in the built environment of village, town, city; in the churches I've visited, the conversations, and the people I've met. I've wanted to find a sense of the presence of God, and know God to be faithful to the creation, so I can hope into the future. I've come to the realization that the emotional roller-coaster I'm on is the new normal, and I need to learn how to work constantly on the despair that overwhelms me at times, to use lament—and particularly the range and depth of the psalms—to turn to a fierce hope that will motivate me to action. How that is to be achieved is the work of poets and prophets, who know what loss and grace are.

In *Gentle Regrets*, Roger Scruton describes how he regained his religion, and writes movingly about loss. He concludes the book with his reflections on the *Jubilate Deo*, Psalm 100. "Once we came before God's presence with a song; now we come before his absence with a sigh," he laments. What might it be like to lose religion, to lose what the Church of England brings to national life? Scruton writes in his final chapter, "Regaining my Religion":

> If you see things in that way you will find it difficult to share the view of Enlightenment thinkers that religious decline is no more than the loss of false beliefs; still less will you be able to accept the postmodernist vision of a world now liberated from absolutes, in which each of us constructs guidelines of his own, and that the only agreement that counts is the agreement to differ. The decline of Christianity, I maintain, involves, for many people, not the freedom from religious need, but the loss of concepts that would enable them to assuage it and, by assuaging it, to open their knowledge and their will to the human reality. For them the loss of religion is an epistemological loss—a loss of knowledge. Losing that knowledge is not a liberation but a fall . . . .
>
> In our civilization, therefore, religion is the force that has enabled us to bear our losses and so to face them as truly ours. The loss of religion makes real loss difficult to bear; hence people begin to flee from loss, to make light of it, or to expel from themselves the feelings that make it inevitable . . . . Modern people pursue not penitence but pleasure, in the hope of achieving a condition in which renunciation is pointless since there is nothing to renounce. Renunciation of love is possible only when you have learned to love. This is why we see emerging a kind of contagious hardness of heart, an assumption on every side that there is no tragedy, no grief, no mourning, for there is nothing to mourn. There is neither love nor happiness—only fun. For us, one might be tempted to suggest, the loss of religion is the loss of loss.
>
> Except that the loss need not occur.[173]

It's a passage that has stayed with me since I first read it. It gives voice to the constant sense of lament that I feel—particularly as I contemplate the natural world around under such pressure, but also as I think about the Church and what it has meant—and could offer yet more—to enable the encounter with what is Other in our lives, and enable humanity to face the drastic future that many are now contemplating and predicting.

To throw yourself into something "like there's no tomorrow!" is to live life to the full. This journey by narrowboat was a personal immersion into different elements and a different way of living—slower, more contemplative, enabling engagement with the environments around with the attention of an amateur naturalist and historian. We went for the quirky, the surprising—which so often became moments of grace, occasions that lifted the soul away from the dark fears and anxieties that are part (though not all) of the normal state of mind for anyone who faces into the inevitable consequences of anthropogenic climate catastrophe.

The penitential psalms

It was a time of greater engagement with the wisdom of the psalms; starting with the penitential ones (6, 32, 38, 51, 102, 130, 143), but also drawing on the emotional and moral wisdom of these ancient hymns of praise to God. I've drawn on other literature too—poetry (including my own), novels, philosophy, theology. Despite the constant chugging, this was a time of silence, such as contemplatives find when they accept the distractions of thought and life and find a deeper place to be and to pray within. Martin Laird describes a retreat where the chainsaw started up next door at two o'clock each day; and how Gareth managed to find the silence in that extreme noise.[174] After a while, you can train yourself not to mind, not to engage with the negative irritation. The chug of a narrowboat engine has its own rhythms and music.

This was a time of moving from south to north, through the different cultural nuances of the UK, during a time of intense unease that was continuously there. I found myself wondering of everyone I encountered how they had voted in June 2016; and one knew they were wondering the same about you. It was often impossible to tell, which stirred up the usual prejudices and the instant judgements we make.

England does change, significantly, from south to north. Poverty is much more evident in the back streets of Burnley or Accrington. It's also there in the upkeep of the canals. Because the affluent don't tend to travel much beyond the Midlands in their narrowboats, there is less money spent on the locks, and the clearance of debris. Once-thriving commercial centres, like Stoke, or Blackburn, now struggle visibly with drugs, and with increased violence and the threat of violence. We needed an anti-vandal key from Wigan onwards. Such poverty and neglect lead to disaffection; to grievances that overlay each other into a compound resentment. Why should climate change be anything to worry about, when you can't manage on less than the living wage, or universal credit doesn't stretch far enough? When heating the home in winter is already too expensive?

Yet still we saw the towpaths used by everyone, their natural wildlife appreciated, offering release, raising the spirits through the free and graceful beauty.

A "meanwhile" time

This was an in-between year, a "meanwhile" time, a *saeculum*, from Autumn 2017 to June 2018. The May and June spent on *The Lark Ascending* provided the opportunity to slow right down and attend to the environment around me in a way I hadn't done for years—perhaps since childhood—and to reflect. Chugging along at three miles an hour gives you plenty of time to focus your attention.

Inwardly, too: I had the chance to think and pray, to contemplate. Martin Laird proved helpful as he describes the power of holding one word in your mind and heart as one prays:

> When the prayer word becomes second nature to us, it becomes more than a spear or a shield or a place of refuge: but an endlessly flowing inner spaciousness in which we "live and move and have our being" (Acts 17:28). Therefore, we do not have to flee from our life circumstances or from our thoughts and feelings (yet we are free to if common sense so dictates). These thoughts and feelings are themselves porous to this spacious flow; they, too, manifest the silence we seek.[175]

I had needed to live in the shadows for a while. The penitential Psalm 130 spoke to this need to wait on the Lord. I imagined, as Elizabeth Cook does in her novel *Lux*, how King David retreated to a dark cave to offer his guilt to God.

> Out of the depths have I cried to you, O Lord;
> Lord, hear my voice;
> let your ears consider well the voice of my supplication.
> If you, Lord, were to mark what is done amiss,
> O Lord, who could stand?
> But there is forgiveness with you,
> so that you shall be feared.
> I wait for the Lord; my soul waits for him;
> in his word is my hope.
> My soul waits for the Lord,
> more than the night watch for the morning,
> more than the night watch for the morning.
> O Israel, wait for the Lord,
> for with the Lord there is mercy;
> With him is plenteous redemption
> and he shall redeem Israel from all their sins.
>
> *Psalm 130*

Being in the dark is hard. It is also the place of the womb. Alves captures the creativity of the shadowland, of darkness:

> And I also love the darkness which abides inside the deep and lovely woods of Frost's poetry, and the light which fractures through unquiet waters in Eliot's poems, and the eerie atmosphere of the gothic cathedral, which reminds me of the entrails of the great fish inside the sea: a sunken cathedral . . . My whole Being reverberates, and I know that it belongs to the darkness of the woods, to the depth of the sea, to the mystery of the cathedral . . . . If lights are turned on I am homeless.[176]

Keller talks of "apophatic entanglement", drawing on Nicholas of Cusa's negative theology:

> Thus Cusa, speaking of the cloud, precipitates a fresh event of the speech that unspeaks itself, of what had been called negative or apophatic theology—from its start a millennium earlier an

> intensively philosophical operation. It was never separable from its contrasting *kataphasis*, its eloquent affirmations. Such a theology performs its negations for the sake of the most positive relations possible.[177]

She means we need to enter into the cloud of unknowing, the darkness, the shadows, so that we may know the true light, light that is not bound by space, the light that does not blind us and make us homeless, but enables us to find ourselves in all the complexity of despair, lament, and fierce hope.

The anxieties I carried were partly, I came to realize, due to my natural tendency to absorb the moods and atmosphere around me, which take over if I'm not careful. I wondered if I had internalized what I came to see as the deep anxieties of the Church of England. I couldn't be that confident, sunny, fun-loving, mission-minded leader that I assumed was required of those who hold senior posts in the Church today. Instead I lived in the shadows, soaking up the doubt, the anxiety, and making it my own. It became memories of things I'd done badly; people I'd hurt; uncomfortable things I knew about myself:

> Why are you so full of heaviness, O my soul,
> and why are you so disquieted within me?
> O put your trust in God;
> for I will yet give him thanks,
> who is the help of my countenance, and my God.
>
> *Psalm 42:6–7*

Living on water gives a different perspective. You're in another element. When you're on board a narrowboat, the foundations below you are fluid. They don't shake, but flow. The deep connections with the world, and with the mystery of God, are felt in a different way. Richard Rohr describes how we are different when we learn to breathe under water:

> Only those who have tried to breathe under water know how important breathing really is, and will never take it for granted again. They are the ones who do not take shipwreck or drowning lightly; they are the ones who can name "healing" correctly, they

> are the ones who know what they have been saved *from*, and the only ones who develop the patience and humility to ask the right questions of God and of themselves.[178]

I learned to take the buffeting of the weather in the face—the sun, the wind—and began to learn the lessons of the water. When we walk through clouds, we breathe under water. We absorb the myriad droplets that drop like mercies unstrained. Keller asks:

> What is a cloud, after all? Not a one, not a fluffy unit, but a collection of billions of water droplets, frozen crystals, each folded around a bit of dust, each utterly distinct. A cloud is a mobile manifold, as are each of us, as are each of our contexts.[179]

Water is the material of the sacrament of baptism. I began to find its regenerative, entangling power, the quality of its blessing, that is twice blest.

The psalms are ancient hymns and poems that hold a mirror to humanity. In these 150 songs there is the whole range of human experience, from joy to lamentation, celebrating the beauty of the natural world, and the tragedy of broken relationship. There are psalms that are full of bitterness, anger, and cursing, which are painful and difficult to read, and those that are sublime, so that one's spirit soars.

The psalms become our own as we pray them and become familiar with them. Our cry of joy or moan of lament is expressed in the depths of the psalms. The words express our yearning today, as Ellen Davis says, *from the inside.*[180]

◆ ◆ ◆

Perhaps birds are particularly beloved of God, as they nest at home at God's altars:

> The sparrow has found her a house
> and the swallow a nest where she may lay her young:
> at your altars, O Lord of hosts, my King and my God.
>
> *Psalm 84:2*

It helped my search for belonging in the Church and in the world as I read psalms that celebrate God's creative energy. I found again the love of the nature God sustains—and also God's desire that humankind might live at harmony with the natural world, and the disturbing wrath of God at wrongdoing.

The Lark Ascending was house and nest for a while, but one which carried us through water. The same psalm speaks of the blessedness of water when life is barren:

> Blessed are those whose strength is in you,
> in whose heart are the highways to Zion,
> Who going through the barren valley find there a spring,
> and the early rains will clothe it with blessing.
>
> *Psalm 84:4–5*

The wisdom of the psalms lends itself to theopoetics. It enables an imaginative and poetic approach to the resources of faith. Their living metaphors "re-describe reality in such surprising ways that we feel touched by them," as Cas J. A. Vos says, in *Theopoetry of the Psalms*.[181] Davis quotes George Steiner, that "A poem is maximal speech," as she writes of preaching the psalms.[182] She says, as poetry, you can't read the psalms in a distracted frame of mind: "You cannot skim them . . . You have to dwell on the words, and the reward for doing so is a fresh view of the world."[183] They take you deep into the human condition, to the bottom of despair. You breathe under water in the psalms, learning to draw near to God, to receive God's blessing from the depths, from the clouds. The inspiration of the psalms is of a different element.

Theopoetics captures that disturbance I experienced, the dissonance of frenetic energy with deep anxiety. We are terrified as a nation, I believe, and we don't know what to do with that terror. The foremost terror is of climate change, out of control. Keller says it's unspeakable. She's right.

> I make reference and will continue to make reference to the climate, the atmosphere in which clouds play their mysterious role in cooling and in warming, as an embodiment of the apophatic entanglement. . . . Even as global warming is becoming common parlance, it remains its own kind of unspeakable: a fright with one constituency and censorship with another.[184]

We face the destruction of the fragile beauty of this world, this creation that is God's gift to us. I feel God's rage at what we have done. That, too, I had internalized—God's wrath—and I didn't know how to live with it.

> For a thousand years in your sight are but as yesterday,
> which passes like a watch in the night.
> You sweep them away like a dream;
> they fade away suddenly like the grass.
> In the morning it is green and flourishes;
> in the evening it is dried up and withered.
> For we consume away in your displeasure;
> we are afraid at your wrathful indignation.
>
> *Psalm 90:4–7*

Church leaders think we can solve the decline of attendance in our churches with heroic action and strategies, attractive branding, and heritage. When really the deep anxiety we all feel is that the world is getting hotter, the seas are warming, and within decades it won't matter any more that our churches are empty.

◆ ◆ ◆

How might it be possible to recapture imagination of the world for God—who cries out to us as we cry out in the psalms—in lament, in pain, in joy and praise, in profound need, in yearning and desire?

Davis writes that the psalmist's foremost state is not self-confidence, but conscious helplessness and utter dependence upon God. Richard Rohr says the same. To be there is deeply uncomfortable, so it's crucial to draw on the psalms as a resource, as people (including Jesus) have done through the ages. They did not just help me personally, as I tried to work out God's call for me. They also speak to the whole of humanity, calling us to account for our abuse of the created order and our lack of stewardship. They express fervent prayer for the poor of the world who suffer most as the world is devastated by climate change.

They offer the words of a lament for the future with a fierce hope.

POSTSCRIPT

Cumbria

Sunday 24 June. Midsummer's Day. My mother, had she lived beyond her sixty-seven years, would have been eighty today. After church, Peter and I head for the hills, and after finding the road closed to Loweswater, come in by Lorton Vale to Buttermere. Lunch of butternut squash and goat's cheese risotto at the Bridge Hotel, then we walk around the lake.

"A given day"

"In Devon, they'd call this day 'given,'" says Peter. "A given day." Warm, a light breeze from the west. People come to enjoy this wonderful spot in different ways. There are those who are swimming, though none we saw are skinny dipping. Others are in the air, suspended on giant kites. The woods give the air that lovely heavy smell. The air is clean. No sign of ash die-back at all on the old trees, covered in healthy growth. The mountain Haystacks is high above—Wainwright's favourite mountain, on which his ashes are scattered. The path leads us through the woodland and through a tunnel, which reminds me of Braunston, Harecastle, Foulridge. We see walls built on the rock; a large Herdwick ram; the foxgloves in the gorse.

Peter and I remember walks we've done here, over the tops. We don't know these mountains as well as the Wasdale ones—so it feels good to anticipate the years of exploration ahead. Mellbreak is there, alongside Crummock Water. Melvyn Bragg's daughter Marie-Elsa, who is a priest, has written *Towards Mellbreak*.[185] It captures the bleak life that's led on these hills, the isolation of the hill farmer as, through the post-war years, they had to put in place the regulations from Westminster, so far away, including the dipping of sheep in organophosphates. She captures the life and death of belonging to these hills that surround us.

Workington

Back in Workington, Peter and I make our way out to the lighthouse. In the evening, we watch a ship reverse into the Prince of Wales Dock, ready to unload timber for the paperboard factory, Iggersund, that's just north of Workington. Then, as we walk, the overwhelming sound of skylark and scent of honeysuckle, combined.

When Viv and I were on the River Nene, we visited Fotheringhay. Mary Queen of Scots was imprisoned there for seventeen years. That was the end of her journey, which began when she was defeated by her rebellious Scottish subjects in 1568. Then she fled to England, landing in Workington. She enjoyed the hospitality of the Curwen family for a few days, before being housed in what was then known as the Warden's Tower at Carlisle Castle. Sir Francis Knollys was appointed to ensure she didn't escape; he allowed her to walk on the grass in front of the castle—thereafter known as "Lady's Walk". Eventually Mary was persuaded to leave for Bolton Castle in Yorkshire, there beginning her southward journey to Fotheringhay, where she was beheaded in 1587.

From Workington, Criffel (a hill on the Scottish coast) is only twenty miles away. We can see the Isle of Man, and the height of Snaefell, as we watch the fishing boats motor out as the tide begins to come in. How I love the sea. Living in a town that's also a port and harbour is a delight.

Renewing his baptism in the deep waters of life

Peter went on his ordination retreat that week. He was in silence. But does WhatsApp count? He sent me messages. "Am breaking the retreat silence to show you scenes of Rydal Hall," he wrote. "Decided to take a swim in Rydal Water just before my interview with the Bishop." It's an extraordinary thing to do in today's world—to give your life, publicly and boldly, to serve others. It's to say, "My life is not my own. It's a gift I have received, and a gift I give to others." When the bishop laid his hands on Peter's head that afternoon, once ordained, always ordained—for he has taken vows that are binding on him for life, vows that are a public statement of his commitment to be as Christ to others.

◆ ◆ ◆

When we consider God's creation, we are only too aware that all is not well. Human destruction and degradation: forests felled, rainforest burning furiously; fossil fuels consumed; carbon dioxide pumped into the atmosphere and into the oceans. We know the polar ice caps are melting; how sea levels are rising; how flooding is now a reality for some, drought and desert—a lack of water—a disaster for others. The world is suffering, and humanity with it.

Pierre Teilhard de Chardin, a French philosopher and Jesuit who trained as a palaeontologist and geologist, died in 1955. Teilhard de Chardin said the entire world is "*le milieu divin*". In the early 1920s he wrote "The Mass on the World"; a devout priest, he was miles from any church or altar on which he might celebrate the Mass. He was in a wild land, far from civilization, but the whole of creation around him seemed to sing with the presence of God. The body of Christ, the blood of Christ were there, to be discerned in the land which surrounded him:

> Since once again, Lord— . . . [now] in the steppes of Asia—I have neither bread, nor wine, nor altar, I will raise myself beyond these symbols, up to the pure majesty of the real itself; I, your priest, will make the whole earth my altar and on it will offer you all the labours and sufferings of the world.[186]

With all its immense heartache and suffering

He imagines everything taken up, with all its immense heartache and suffering, as the bread and wine are lifted. As Teilhard, the priest, holds up the chalice and paten of the world around him, all the pain and joy of the created order is there, as it groans inwardly to be one with God:

> This bread, our toil, is of itself, I know, but an immense fragmentation; this wine, our pain, is no more, I know, than a draught that dissolves. Yet in the very depths of this formless mass you have implanted—and this I am sure of, for I sense it—a desire, irresistible, hallowing, which makes us cry out, believer and unbeliever alike: "Lord, make us *one*."[187]

Teilhard believed that there is an underlying unity to all things, sustained by the love of God which spends itself again and again for the creation. That sacrificial love is at the heart of things; it is the "Mass on the world".

The world is sacramental

The world is infused with the creative love of God. More than that, it is through suffering and self-sacrifice that this love gives life. God emptied himself, became human, became matter, that matter might be reconciled. "Through your own incarnation, my God, all matter is henceforth incarnate," writes Teilhard.[188] The living, dying, and rising again of Jesus Christ is an action that is continually repeated within creation, within each of us. There is a larger story going on, a story that tells us death is never stronger than love. Like the yeast that leavens the bread, that ferments the wine, but which dies in the process, the sacrificial love of God in Christ gives life.

We see life, the fruit of the Spirit, all around us. In 1962, Rachel Carson predicted a "silent spring", as humanity continued to exploit the natural world.[189] In fact, nature provides a raucous summer, given half the chance. Forests regenerate; species return; nature recovers. When we, the human race, respond to God in love and respect for the integrity of creation, with a much stronger sense of stewardship and responsibility, we find ourselves, not pilgrims through an increasingly barren land, but joyful pilgrims, rather, in a rich and abundant world, the divine milieu, which reflects the love and glory at the heart of God.

Our own lives become richer too: relationships strengthened in peace and love; a sense of purpose and meaning in life; inner joy; a wild patience that lives in hope.

As we receive the cup of salvation, the bread of life, we participate in a "Mass on the world" which brings hope to the whole created order. We experience the release from helplessness and despair, we see a glory about to be revealed. Our desire, as those who consume the sacrament of Christ's body, is for union with God beyond life, beyond death. This union with God belongs to the whole of creation, of which we are all but parts, with our own role to play, fulfilling what we are meant to be within God's ordered universe, now so disordered by human sin.

This is my body. This is my blood. We receive Christ, remembering that he goes to his death. His self-sacrifice lies at the heart of the love of God. This is a God in Christ who gives, and gives, and gives again in order that we and all creation might have life. Let us not take, and take, and take in return, but grow in self-sacrificial love ourselves, for the sake of God's creation.

"A vast similitude interlocks all"

The week of the ordination began with a walk around Buttermere. It ends with Peter and me swimming in Loweswater. The water is delicious; the scenery stunning. We walk there from Fangs Brow, noticing the wayside flowers: the elder, just going over towards its blood-red berry; the thistle, reminding us that Scotland isn't far away; rosebay willowherb in full song; an umbellifer that isn't cow parsley (which is finished now), so let's call it Queen Anne's lace; meadow vetch; clover in a meadow, and harebells reflecting the sky. Nor should we forget the humble nettle and dock in flower.

The walk takes us along the lake on the wooded south side: through Hudson's Place, with its parquetry dating from 1741, and hay all gathered in; alongside ancient hedgerows, where the roots grip the side of the dry earth, and untidy branches grow away from the wind—hawthorn again, now tired of the heat, no longer singing the song of May. We clamber over double stiles, down to the lake, and find a quiet spot in the woods to skinny dip.

Walt Whitman comes and goes, as my thoughts sink away from words, floating on my back, buoyed by the water:

> On the beach at night alone,
> As the old mother sways her to and fro singing her husky song,
> As I watch the bright stars shining, I think a thought of the clef
> of the universes and of the future.
> A vast similitude interlocks all, . . .
> All distances of place however wide,
> All distances of time, . . .
> All souls, all living bodies though they be ever so different.[190]

I close my eyes, and the sun shines red and gold through my eyelids. I feel myself turning in the drift and wonder if I imagine it—the water taking me, and turning me, as a boat turns with the tide. I am one in its element; taking its risk, its weight, its potential and power. My frailty taken in its depths.

> Out of the depths I cry to you, O Lord.

Bibliography

Alves, Rubem A., *The Poet, The Warrior, The Prophet* (SCM Press, [1990], 2002).

Armitage, Simon, *Walking Home: Travels with a Troubadour on the Pennine Way* (Faber & Faber, 2013).

Augustine, *Confessions* (Penguin, 1961, trans. R. S. Pine-Coffin).

Bendell, Jem, "Deep Adaptation: A Map for Navigating Climate Tragedy", IFLAS Occasional Paper 2, 27 July 2018; <https://www.lifeworth.com/deepadaptation.pdf>.

Bragg, Marie-Elsa, *Towards Mellbreak* (Chatto & Windus, 2017).

Carson, Rachel, *Silent Spring* (Penguin Classics, [1962] 2000).

Coakley, Sarah, *God, Sexuality, and the Self: An Essay 'On the Trinity'* (Cambridge University Press, 2013).

Cocker, Mark, and Mabey, Richard, *Birds Britannica* (Chatto & Windus, 2005).

Cocker, Mark, *Our Place: Can We Save Britain's Wildlife Before it is Too Late?* (Jonathan Cape, 2018).

Colwell, Mary, *Curlew Moon* (William Collins, 2018).

Conway Morris, Simon, *Life's Solution: Inevitable Humans in a Lonely Universe* (Cambridge University Press, 2003).

Cook, Elizabeth, *Lux* (Scribe Publications, 2019).

Crawford, Matthew, *The World Beyond Your Head: How to Flourish in an Age of Distraction* (Penguin Random House, 2015).

Davis, Ellen F., *Proverbs, Ecclesiastes, and the Song of Songs* (Westminster John Knox Press, 2000).

Davis, Ellen F., *Wondrous Depth: Preaching the Old Testament* (Westminster John Knox Press, 2005).

Deane-Drummond, Celia, and Artinian-Kaiser, Rebecca (eds), *Theology and Ecology Across the Disciplines: On Care for Our Common Home* (T&T Clark, 2018).

Ford, Martin, *Rise of the Robots: Technology and the Threat of a Jobless Future* (Basic Books, 2015).
Gaw, Matt, *The Pull of the River: A Journey into the Wild and Watery Heart of Britain* (Elliott & Thompson, 2018).
Goodhart, David, *The Road to Somewhere: The Populist Revolt and the Future of Politics* (C. Hurst & Co., 2017).
Greer, Germaine, *White Beech: The Rainforest Years* (Bloomsbury, 2014).
Harris, Alexandra, *Weatherland: Writers and Artists under English Skies* (Thames & Hudson, pbk, 2016).
Helm, Dieter, *Green and Prosperous Land: A Blueprint for Rescuing the British Countryside* (William Collins, 2019).
Hoskins, W. G., *The Making of the English Landscape* (Hodder and Stoughton [1955], Penguin reprint, 1986).
Hunter-Blair, Andrew, *Fenland Waterways: A Map and Commentary on the Waterways of the Middle Level* (Imray, Laurie, Norie & Wilson Ltd, 2016).
Keefe-Perry, L. Callid, *Way to Water: A Theopoetics Primer* (Cascade Books, 2014).
Keller, Catherine, *Cloud of the Impossible: Negative Theology and Planetary Entanglement* (Columbia University Press, 2014).
Kurlansky, Mark, *Salt: A World History* (Vintage, 2003).
Laird, Martin, *A Sunlit Absence: Silence, Awareness, and Contemplation* (Oxford University Press, 2011).
Lanchester, John, *The Wall* (Faber & Faber, 2019).
Mabey, Richard, *Weeds: How Vagabond Plants Gatecrashed Civilisation and Changed the Way we Think about Nature* (Profile Books, 2010).
Macaulay, Rose, *The World My Wilderness* (The Book Club, 1950).
Macfarlane, Robert, *The Wild Places* (Granta, [2007] 2017).
Macfarlane, Robert, *The Old Ways: A Journey on Foot* (Hamish Hamilton, 2012).
Macfarlane, Robert and Morris, Jackie, *The Lost Words: A Spell Book* (Hamish Hamilton, 2017).
McCann, Justin, OB (ed.), *The Cloud of Unknowing* (Burns Oates, 1952).
McCarthy, Michael, *The Moth Snowstorm: Nature and Joy* (John Murray, 2016).

McGilchrist, Iain, *The Master and his Emissary: The Divided Brain and the Making of the Western World* (Yale University Press, 2009).

Mirrlees, Hope, *Lud-in-the-Mist* (Gollancz, [1926] 2018).

Monbiot, George, *Feral: Searching for Enchantment on the Frontiers of Rewilding* (Allen Lane, 2013).

Northcott, Michael S., *A Political Theology of Climate Change* (SPCK, 2014).

Orwell, George, *The Road to Wigan Pier* (Penguin Classics, [1937] 2001).

Peers, E. Allison, *Mother of Carmel: A Portrait of St Teresa of Jesus* (SCM Press, 1945).

Rackham, Oliver, *Woodlands* (William Collins, 2015).

Rohr, Richard, *Breathing Under Water: Spirituality and the Twelve Steps* (SPCK, 2016).

Sakimoto, Philip J. "Prologue: Understanding the Science of Climate Change", in Celia Deane-Drummond and Rebecca Artinian-Kaiser (eds), *Theology and Ecology Across the Disciplines: On Care for Our Common Home* (T&T Clark, 2018, pp. 7–22).

Sayers, Dorothy L., *The Nine Tailors* (Hodder & Stoughton, [1934] 2003).

Scruton, Roger, *Gentle Regrets: Thoughts from a Life* (Continuum, 2005).

Swift, Graham, *Waterland* (Picador, 2010).

Teilhard de Chardin, Pierre, *Hymn of the Universe* (William Collins, [1961] 1970).

Thompson, Flora, *Lark Rise to Candleford* ([1939] Oxford University Press, The World's Classics, 1975).

Uglow, Jenny, *Mr Lear: A Life of Art and Nonsense* (Faber & Faber, 2017).

Vos, Cas J. A., *Theopoetry of the Psalms* (T&T Clark, 2005).

Wallace-Wells, David, *The Uninhabitable Earth: A Story of the Future* (Allen Lane, 2019).

Ward, Frances and Sudworth, Richard (eds), *Holy Attention: Preaching in Today's Church* (Canterbury Press, 2019).

White, Lynn Jr., "The Historical Roots of Our Ecologic Crisis", *Science* 155 (1967), pp. 1203–7.

Notes

[1] See Elizabeth Cook, *Lux* (Scribe Publications, 2019), a novel which explores the penitential psalms of David, and how Thomas Wyatt came to translate them (to "English them") in the 1530s.

[2] <https://rebellion.earth/the-truth/about-us/>

[3] David Wallace-Wells, *The Uninhabitable Earth: A Story of the Future* (Allen Lane, 2019).

[4] Jem Bendell, "Deep Adaptation: A Map for Navigating Climate Tragedy", IFLAS Occasional Paper 2, Institute for Leadership and Sustainability, University of Cumbria, 27 July 2018, <https://www.lifeworth.com/deepadaptation.pdf>.

[5] <https://www.churchtimes.co.uk/theology-slam>

[6] For a Podcast of Hannah Malcolm's talk see <https://www.churchtimes.co.uk/articles/2019/8-march/regulars/podcast/podcast-theology-slam-final-hannah-barr-sara-prats-hannah-malcolm>; and the article can be found here <https://www.churchtimes.co.uk/articles/2019/15-march/comment/opinion/climate-chaos-and-collective-grief>.

[7] Philip J. Sakimoto, "Prologue: Understanding the Science of Climate Change", in Celia Deane-Drummond and Rebecca Artinian-Kaiser (eds.), *Theology and Ecology Across the Disciplines: On Care for Our Common Home* (T&T Clark, 2018), pp. 7–22.

[8] Sakimoto, "Prologue", p. 9.

[9] Sakimoto, "Prologue", p. 13.

[10] Sakimoto, "Prologue", p. 17.

[11] Sakimoto cites Princeton University, "As global temperatures rise, children must be central to climate change debates": *ScienceDaily* (2016), <https://www.sciencedaily.com/releases/2016/05/160504121330.htm>.

[12] Sakimoto, "Prologue", p. 21, his emphasis.

[13] Rowan Williams, private email correspondence with author.

[14] Williams, *ibid.*.

[15] Bendell, "Deep Adaptation", p. 8.

[16] Bendell, "Deep Adaptation", p. 24.

[17] <http://w2.vatican.va/content/francesco/en/encyclicals/documents/papa-francesco_20150524_enciclica-laudato-si.html>.

[18] Bendell, "Deep Adaptation", p. 12.

[19] Bendell, "Deep Adaptation", p. 22.

[20] Bendell, "Deep Adaptation", p. 23.

[21] <https://jembendell.com/category/deep-adaptation/>.

[22] <https://knepp.co.uk/>; see Isabel Tree, Wilding: The Return of Nature to a British Farm (Pan McMillan, 2019)

[23] L. T. C. Rolt, *Narrow Boat* (The History Press, [1944] 2014).

[24] <https://www.waterways.org.uk/>.

[25] Dieter Helm, *Green and Prosperous Land: A Blueprint for Rescuing the British Countryside* (William Collins, 2019).

[26] Bendell, "Deep Adaptation", p. 25.

[27] <https://www.foxboats.co.uk/>.

[28] From George Meredith, "The Lark Ascending". The poem can be read in full at <https://allpoetry.com/The-Lark-Ascending>.

[29] From John Clare, "The Skylark". The poem can be read in full at <https://www.poetryfoundation.org/poems/43950/the-skylark>.

[30] <https://www.bbc.co.uk/news/science-environment-49134175>.

[31] L. Callid Keefe-Perry, *Way to Water: A Theopoetics Primer* (Cascade Books, 2014), p. 49.

[32] Keefe-Perry, *Way to Water*, p. 49; Rubem A. Alves, *The Poet, The Warrior, The Prophet* (SCM Press, [1990] 2002).

[33] Iain McGilchrist, *The Master and his Emissary: The Divided Brain and the Making of the Western World* (Yale University Press, 2009).

[34] McGilchrist, *The Master and his Emissary*, pp. 428ff.

[35] McGilchrist, *The Master and his Emissary*, pp. 433–4.

[36] See Prayer 6 at <http://justus.anglican.org/~ss/commonworship/hc/preptable.html>.

[37] Augustine, *Confessions* (Penguin, 1961, trans. R. S. Pine-Coffin), Book X, Chapter 6, pp. 211–12.

[38] <https://www.nhm.ac.uk/discover/news/2019/february/whales-and-dolphins-are-getting-stuck-in-fishing-nets-around-the.html>.

[39] Martin Ford, *The Rise of the Robots: Technology and the Threat of a Jobless Future* (Basic Books, 2015), p. xvii.

40 <https://www.economist.com/leaders/2019/02/09/the-truth-about-big-oil-and-climate-change>.

41 <https://www.ipcc.ch/>.

42 Bendell, "Deep Adaptation", p. 24.

43 See her TED talk at <https://www.ted.com/speakers/greta_thunberg>.

44 Schrödinger used the German word "*Verschränkung*" but translated it himself as "entanglement".

45 Catherine Keller, *Cloud of the Impossible: Negative Theology and Planetary Entanglement*, (Columbia University Press, 2015), p. 5.

46 Keller, *Cloud of the Impossible*, p. 137, her emphasis.

47 Flora Thompson, *Lark Rise to Candleford* ([1939] The World's Classics, 1975).

48 Helm, *Green and Prosperous Land*. See also the website <http://www.dieterhelm.co.uk/natural-capital/environment/soils-policy-and-the-25-year-environment-plan/>.

49 Robert Macfarlane and Jackie Morris, *The Lost Words: A Spell Book* (Hamish Hamilton, 2017).

50 Mark Cocker, *Our Place: Can We Save Britain's Wildlife Before it is Too Late?* (Jonathan Cape, 2018).

51 <https://www.rspb.org.uk/get-involved/campaigning/let-nature-sing/>

52 <https://www.rspb.org.uk/birds-and-wildlife/advice/how-you-can-help-birds/where-have-all-the-birds-gone/is-the-number-of-birds-in-decline/> (accessed 28/11/2019)

53 Alves, *The Poet, the Prophet, the Warrior*, p. 7.

54 Graham Swift, *Waterland* (Picador, 2010), pp. 20–21, his emphasis.

55 Dorothy L. Sayers, *The Nine Tailors* (Hodder & Stoughton, [1934] 2003), p. 187.

56 Sayers, *The Nine Tailors*, p. 369.

57 Andrew Hunter-Blair, *Fenland Waterways: A Map and Commentary on the waterways of the Middle Level* (Imray, Laurie, Norie & Wilson Ltd, 2016), p. 8.

58 Helm, *Green and Prosperous Land*, p. 82.

59 William Wordsworth, *The Prelude*, Book Fourth, lines 247–64. See the whole poem at <http://triggs.djvu.org/djvu-editions.com/WORDSWORTH/PRELUDE1850/Prelude1850.pdf>.

60 Teresa of Avila, from *The Life*, quoted in E. Allison Peers, *Mother of Carmel: A Portrait of St Teresa of Jesus* (SCM Press, 1945), p. 31.

61 Justin McCann, OB (ed.), *The Cloud of Unknowing* (Burns & Oates, 1952), p. 12.

62 Alves, *The Poet, The Warrior, The Prophet*, p. 16–17.

[63] Sarah Coakley, *God, Sexuality, and the Self: An Essay 'On the Trinity'* (Cambridge University Press, 2013), p. 26.

[64] Robert Macfarlane, *The Old Ways: A Journey on Foot* (Hamish Hamilton, 2012), p. 132.

[65] Alves, *The Poet, The Warrior, The Prophet*, p. 24.

[66] <https://news.nationalgeographic.com/2018/06/wildlife-watch-eel-smuggling-operation-broken-glass-maine/>.

[67] Swift, *Waterland*, p. 205.

[68] Robert Macfarlane, *The Wild Places* (Granta, [2007] 2017), pp. 1–2.

[69] From Robert Frost's "Birches"; the poem can be read in full at <https://www.poetryfoundation.org/poems/44260/birches>.

[70] See Simon's blog at <https://nbsg.wordpress.com/>. Matthew Arnold's poem "The Scholar-Gypsy" can be found at <https://www.poetryfoundation.org/poems/43606/the-scholar-gipsy>.

[71] Jenny Uglow, *Mr Lear: A Life of Art and Nonsense* (Faber & Faber, 2017).

[72] Alexandra Harris, *Weatherland: Writers and Artists under English Skies* (Thames & Hudson, pbk. 2016).

[73] John Bayley, *Elegy for Iris* (Saint Martin's Press, 1999).

[74] Richard Rohr, *Breathing Under Water: Spirituality and the Twelve Steps* (SPCK, 2016), p. xxi.

[75] Matt Gaw, *The Pull of the River: A Journey into the Wild and Watery Heart of Britain* (Elliott & Thompson, 2018)

[76] Gaw, *The Pull of the River*, p. 64.

[77] Gaw, *The Pull of the River*, p. 65.

[78] Gaw, *The Pull of the River*, p. 73.

[79] Gaw, *The Pull of the River*, pp. 73–4.

[80] Gaw, *The Pull of the River*, p. 75.

[81] Quoted in Gaw, *The Pull of the River*, p. 80.

[82] Gerard Manley Hopkins, last verse of his 1881 poem *Inversnaid*; the poem can be read in full at <http://www.poetrybyheart.org.uk/poems/inversnaid/>.

[83] Ellen F. Davis, *Proverbs, Ecclesiastes, and the Song of Songs* (Westminster John Knox Press, 2000), p. 45, her emphasis.

[84] Keller, *Cloud of the Impossible*, p. 283.

[85] Lynn White, Jr., "The Historical Roots of Our Ecologic Crisis", in *Science* 155 (1967), pp. 1203–7. White's thesis has been soundly refuted by Michael S. Northcott: *A Political Theology of Climate Change* (SPCK, 2014), p. 106.

[86] See Northcott, *A Political Theology of Climate Change*.

[87] Simon Conway Morris, *Life's Solution: Inevitable Humans in a Lonely Universe* (Cambridge University Press, 2003).

[88] Conway Morris, *Life's Solution*, p. 313, quoting G. K. Chesterton, "The Blue Cross", the first story in *The Innocence of Father Brown* (1940), p. 22.

[89] Donne's sermon of 29 January 1625 [1626], quoted in Ellen F. Davis, *Wondrous Depth: Preaching the Old Testament* (Westminster John Knox Press, 2005), p. 61.

[90] Michael McCarthy, *The Moth Snowstorm: Nature and Joy* (John Murray, 2016), pp. 153–4 (abridged).

[91] Helm, *Green and Prosperous Land*, p. 43.

[92] Helm, *Green and Prosperous Land*, p. 196.

[93] Cocker, *Our Place*, pp. 182–3.

[94] Alves, *The Poet, The Warrior, The Prophet*, p. 101.

[95] See the image at <http://www.musee-rodin.fr/fr/collections/sculptures/la-cathedrale>.

[96] Cook, *Lux*, 2019.

[97] Mark Cocker and Richard Mabey, *Birds Britannica* (Chatto & Windus, 2005), p. 7.

[98] Sayers, *The Nine Tailors*, p. v.

[99] <https://en.wikipedia.org/wiki/Becoming_Jane>. The film is widely available; the scene takes place in the library.

[100] <https://www.oxford.anglican.org/mission-ministry/environment/resources/swifts-churches/>.

[101] Helm, *Green and Prosperous Land,* p. 87.

[102] Cocker, *Our Place*, pp. 114–5.

[103] Cocker, *Our Place*, p. 112.

[104] John Lanchester, *The Wall* (Faber & Faber, 2019).

[105] Marge Piercy, *Body of Glass* (Penguin, 1991).

[106] Cocker, *Our Place*, p. 199.

[107] Mary Colwell, *Curlew Moon* (William Collins, 2018), p. 178.

[108] Cocker, *Our Place*, p. 195.

[109] Germaine Greer, *White Beech: The Rainforest Years* (Bloomsbury, 2014).

[110] Tim Coughlan, "Arthur Bray obituary", The Guardian, 2 January 1999, online at <https://www.theguardian.com/news/1999/jan/02/guardianobituaries1>.

[111] *ibid*.

[112] W. G. Hoskins, *The Making of the English Landscape* (Penguin, 1986), p. 151.

[113] Hope Mirrlees, *Lud-in-the-Mist* (Gollancz, [1926] 2018), p. 47.

[114] Malcolm Guite, "Poet's Corner", *Church Times*, 31 August 2018, back page.

[115] See "The global tree restoration potential", in *Science* Vol. 365 Issue 6448 (5 July 2019), pp. 76–9 DOI: 10.1126/science.aax0848, and article on this report by *The Guardian*'s Environment Editor, 4 July 2019, at <https://www.theguardian.com/environment/2019/jul/04/planting-billions-trees-best-tackle-climate-crisis-scientists-canopy-emissions>.

[116] Cocker, *Our Place*, pp. 285–6.

[117] Cocker, *Our Place*, p. 287.

[118] <https://www.nationaltrust.org.uk/shugborough-estate>.

[119] <https://www.abc.net.au/news/2018-05-20/royal-wedding-meghan-markle-upstaged-by-reverend-michael-curry/9779990>.

[120] See Frances Ward and Richard Sudworth (eds), *Holy Attention: Preaching in Today's Church* (Canterbury Press, 2019).

[121] Matthew Crawford, *The World Beyond Your Head: How to Flourish in an Age of Distraction* (Penguin Random House, 2015).

[122] The UK government Committee on Climate Change recommendation, May 2019.

[123] <https://www.freecycle.org/>

[124] <https://www.ilovefreegle.org/>

[125] <https://www.triodos.co.uk/>

[126] Mark Kurlansky, *Salt: A World History* (Vintage, 2003), pp. 193–4.

[127] Rose Macaulay, *The World My Wilderness* (The Book Club, 1950).

[128] Richard Mabey, *Weeds: How Vagabond Plants Gatecrashed Civilisation and Changed the Way We Think about Nature* (Profile Books, 2010), pp. 214ff.

[129] Macaulay, *The World my Wilderness*, p. 87.

[130] <https://www.markcazalet.co.uk/>.

[131] A suicide bombing attack at the end of a pop concert on 22 May 2017, in which twenty-two people died and many were injured.

[132] The IRA detonated an enormous lorry bomb in the centre of Manchester on 15 June 1996, injuring 200.

[133] <https://www.flickr.com/photos/themakerfieldrambler/15582099932>.

[134] the Orwell text is available on <https://www.orwellfoundation.com/the-orwell-foundation/orwell/essays-and-other-works/the-moon-under-water>.

[135] George Orwell, *The Road to Wigan Pier* (Penguin Classics, [1937] 2001).

[136] <https://www.theguardian.com/books/2011/feb/20/orwell-wigan-pier-75-years>.

[137] Helm, *Green and Prosperous Land*, p. 159.

[138] Helm, *Green and Prosperous Land*, p. 170.

[139] Helm, *Green and Prosperous Land*, p. 225.

[140] Helm, *Green and Prosperous Land*, p. 236.

[141] Helm, *Green and Prosperous Land*, pp. 263, 265.

[142] Helm, *Green and Prosperous Land*, p. xii.

[143] Helm, *Green and Prosperous Land*, p. 10; see also Chapter 9 for his development of this "polluter-pays" principle.

[144] See a digitised booklet about the exchange and the statue at <https://archive.org/details/manchesterabraha00hour>.

[145] Colwell, *Curlew Moon*, p. 35.

[146] Oliver Rackham, *Woodlands* (William Collins, 2015), p. 61.

[147] Andrew Davison, "This was an act of magnificence", *Church Times*, 19 July 2019, p. 18.

[148] <https://www.churchtimes.co.uk/articles/2019/14-june/news/uk/green-health-how-gardens-make-you-sane>.

[149] <https://www.dorotheamackellar.com.au/archive/mycountry.htm>.

[150] Rumer Godden, *The Greengage Summer* (MacMillan General Books,[1958] 1995).

[151] Colwell, *Curlew Moon*, p. 5.

[152] Colwell, *Curlew Moon*, p. 49.

[153] Colwell, *Curlew Moon*, p. 51.

[154] Helm, *Green and Prosperous Land*, p. 184.

[155] <http://kathleenferrier.org.uk/>

[156] Macfarlane, *The Old Ways*, pp. 11ff.

[157] Macfarlane, *The Old Ways*, p. 27.

[158] David Goodhart, *The Road to Somewhere: The Populist Revolt and the Future of Politics* (C. Hurst and Co., 2017).

[159] Davis, *Wondrous Depth*, p. 147.

[160] John Buchan, *Witch Wood* (Hodder & Stoughton, 1927).

[161] Terry Castle, *The Female Thermometer: Eighteenth-Century Culture and the Invention of the Uncanny* (Oxford University Press, 1995).

[162] McGilchrist, *The Master and his Emissary*, p. 350.

[163] . . . or heading south, if you're Simon Armitage walking home from Scotland, as he describes in his excellent *Walking Home: Travels with a Troubadour on the Pennine Way* (Faber & Faber, 2013).

[164] <https://allpoetry.com/Wedding-Wind>

[165] L. M. Montgomery, *Emily of New Moon* (Bantam, [1923] 1980), pp. 13ff.

[166] Harris, *Weatherland*, (Thames and Hudson, pbk, 2016) p. 14.

[167] From Samuel Taylor Coleridge, "Fancy *In Nubibus*, or The Poet In The Clouds"; the poem can be read in full at <https://www.poemhunter.com/poem/fancy-in-nubibus-or-the-poet-in-the-clouds/>.

[168] Harris, *Weatherland*, p. 330.

[169] Hugh Massingham, Introduction to Thompson, *Lark Rise to Candleford*, p. ix.

[170] George Monbiot, *Feral: Searching for Enchantment on the Frontiers of Rewilding* (Allen Lane, 2013).

[171] Thompson, *Lark Rise to Candleford*, p. 22/3.

[172] Rohr, *Breathing Under Water*.

[173] Roger Scruton, *Gentle Regrets: Thoughts from a Life* (Continuum, 2005), pp. 227, 229.

[174] Martin Laird, *A Sunlit Absence: Silence, Awareness, and Contemplation* (Oxford University Press, 2011), pp. 49–54.

[175] Laird, *A Sunlit Absence*, p. 23.

[176] Alves, *The Poet, The Warrior, The Prophet*, p. 8–9.

[177] Keller, *Cloud of the Impossible*, pp. 2–3.

[178] Rohr, *Breathing Under Water*, pp. 123–4.

[179] Keller, *Cloud of the Impossible*, p. 22.

[180] Davis, *Wondrous Depth*, p. 20, her emphasis.

[181] Cas J. A. Vos, *Theopoetry of the Psalms* (T&T Clark, 2005), p. 321.

[182] Davis, *Wondrous Depth*, p. 23.

[183] Davis, *Wondrous Depth*, p. 24.

[184] Keller, *Cloud of the Impossible*, p. 25.

[185] Marie-Elsa Bragg, *Towards Mellbreak* (Chatto & Windus, 2017).

[186] Pierre Teilhard de Chardin, *Hymn of the Universe*, (William Collins, [1961] 1970), p. 19.

[187] Teilhard de Chardin, *Hymn of the Universe*, p. 20.

[188] Teilhard de Chardin, *Hymn of the Universe*, p. 23.

[189] Rachel Carson, *Silent Spring* (Penguin Classics, [1962] 2000).

[190] From Walt Whitman, "On the Beach at Night Alone"; the poem can be read in full at <https://www.poetryfoundation.org/poems/48856/on-the-beach-at-night-alone>.

CPSIA information can be obtained
at www.ICGtesting.com
Printed in the USA
LVHW042249020320
648721LV00005B/526

9 781789 590883